Lecture Notes in Control and Information Sciences

Edited by M. Thoma and A. Wyner

Lecture Notes in Control and Information Sciences

Edited by M. Thoma and A. Wyner

168

P. Dorato, L. Fortuna,
G. Muscato

Robust Control for
Unstructured Perturbations –
An Introduction

Springer-Verlag
Berlin Heidelberg GmbH

Authors

Prof. Peter Dorato
Dept. of Electrical and Computer Eng.,
University of New Mexico
Albuquerque, NM 87131
USA

Prof. Luigi Fortuna
G. Muscato
Dipartimento Elettrico Elettronico
e Sistemistico
Università di Catania
viale A. Doria 6
95125 Catania
Italy

ISBN 978-3-540-54920-8   ISBN 978-3-540-46510-2 (eBook)
DOI 10.1007/978-3-540-46510-2

Typesetting: Camera ready by author
61/3020-5 4 3 2 1 0 Printed on acid-free paper.

PREFACE

These lecture notes are based on lectures given by the first author at the University of Catania, June 4 - 13 , 1990. The lectures focused on the synthesis of robust controllers for feedback systems in the presence of unstructured plant transfer function perturbations.

In part I interpolation theory is used to solve various single-input-single-output (SISO) robust control problems. While the interpolation approach is awkward for multivariable systems, it provides a very natural and simple approach for SISO systems. In particular the interpolation approach requires only elementary knowledge of complex variables, and provides a great deal of physical insight into various robust control problems. The required interpolation theory is developed in some detail.

Part II is devoted to multivariable systems. Two approaches are outlined : the Hankel-norm approach and the two-Riccati-equation approach. In this part only a limited number of results are proven. However MATLAB™ software is presented for numerical solutions.

Most of the papers cited in these lecture notes are included in the IEEE Press reprint volumes ROBUST CONTROL and RECENT ADVANCES IN ROBUST CONTROL.

INDEX

Part I Single Input Single Output systems

<u>Chapter</u> <u>1</u> Introduction

<u>Chapter</u> <u>2</u> Robust Stabilization

<u>Chapter</u> <u>3</u> Nevanlinna-Pick Interpolation Theory

ACRONYMS

BIBO   – Bounded-input-bounded-output.

BR   – Bounded real.

EP   – Exactly proper.

LQG   – Linear-quadratic-Gaussian.

LQR   – Linear-quadratic-regulator.

PA   – Positive analytic.

p.i.p.   – Parity interlacing property.

PR   – Positive real.

RHP   – Right-half-plane.

SBR   – Strictly bounded real.

SPR   – Strictly positive real.

COMMON SYMBOLS

$\mathbb{R}^{n \times m}$ – Space of n x m matrices with real entries.

X^T – Transpose of matrix X.

$\bar{X}$ – Complex conjugate of matrix X.

$\|X\|$ – Norm of matrix X.

$\sigma(X)$ – Singular value of matrix X.

$|X|$ – Determinant of matrix X.

$\rho(X)$ – Spectral radius of matrix X.

$G^*(s)$ – $G^T(-s)$.

$\|F(s)\|_\infty$ – H^∞ norm of function F(s).

H^∞ – Hardy space with bounded H^∞ norm.

$M(H^\infty)$ – Space of matrices with H^∞ entries.

$\|F(s)\|_H$ – Hankel norm of function F(s).

$\sup \Phi(\omega)$ – Limit superior of function $\Phi(\omega)$.

Chapter 1

INTRODUCTION

1.1 A BRIEF HISTORICAL PERSPECTIVE

ROBUST CONTROL is generally defined as the control of uncertain systems with fixed controllers. The term appeared in the literature for the first time in the early seventies, although the basic problem had been studied for many years. In 1927 H. S. Black [4] proposed the use of feedback and large loop gains to reduce sensitivity to plant perturbations. In 1932, with his now classical stability criterion, Nyquist [31] presented a simple frequency domain criterion to determine the stability of a feedback systems in terms of its loop gain. The Nyquist theory dictated how large the loop gain could possibly be if closed-loop stability was to be achieved. Bode [5] combined these two basic results to develop a theory of "robust" system design which dominated the field until the early eighties. Then a number of theories appeared based on Hardy-space concepts and interpolation theory, most notably the result of Tannenbaum [35], Zames and Francis [45], and Kimura [27], which launched a new approach to the synthesis of feedback systems. These results were generalized to multivariable systems through the works of Vidyasagar and Kimura [38], Chang and Pearson [7], Glover [23], and others. Many of the results of this period are collected in the reprint volume ROBUST CONTROL [13]. Another major approach to robust control developed during this period of time is centered on the results of Kharitonov [26] on the stability of interval polynomials. We do not pursue this approach here because to date, the results have been limited to the analysis of uncertain systems rather than synthesis. We wish to focus here on problems and methods that lead to analytical synthesis techniques. Some recent results in robust control theory may be found in the reprint volume RECENT RESULTS IN ROBUST CONTROL [17]. An especially important recent result is the two-Riccati equation approach to the solution of multivariable robust control problem given by Doyle, Glover, Khargonekar, and Francis [20].

1.2 PREREQUISITES

It is assumed that the reader is familiar with the key linear systems concepts of state-space and transfer-function system descriptions, controllability, observability and stability, as developed for example in such standard texts as Chen [8] and Kailath [25]. It is further assumed that the reader has some basic knowledge of functions of a complex variable and matrices, in addition to knowledge of basic of automatic control concepts including such concepts as feedback control, bounded-input-bounded-output (BIBO) stability, Nyquist criterion, sensitivity function, gain and phase margins, and optimal linear-quadratic-regulation (LQR). Most of the above topics are covered in standard introductory texts on automatic control. The LQR problem, and the associated linear-quadratic-Gaussian (LQG) problem are discussed in detail in the books of Kwakernaak and Sivan [28] and Anderson and Mocre [1].

1.3 MODELLING OF UNCERTAIN SYSTEMS AND THE ROBUST CONTROL PROBLEM

Time domain UNSTRUCTURED uncertainty

$$\begin{cases} \dot{x} = (A_0 + \delta A) x + (B_0 + \delta B) u \\ y = (C_0 + \delta C) x \end{cases} \tag{1.1}$$

In this case the data given for Robust control are:

- Nominal A_0 , B_0 , C_0;
- Bounds on perturbation δA , δB , δC , i.e. $\|\delta A\| \leq a$, etc.
  where the symbol $\|\cdot\|$ denotes the norm of a matrix.

Time domain STRUCTURED uncertainty

In this case we have more information about the structure of uncertainty, for example

$$\delta A = g_1 A_1 + g_2 A_2 + \ldots + g_n A_n \tag{1.2}$$

The data given for Robust Control are

- Nominal $A_0, A_1, \ldots\ldots, A_n$; $B_0, B_1, \ldots\ldots B_m$; $C_0, C_1, \ldots, C_p$;
- Bounds on g_i , i.e. $\underline{g}_i \leq g_i \leq \bar{g}_i$.

Frequency domain UNSTRUCTURED uncertainty

ADDITIVE PERTURBATION $\qquad\qquad\qquad$ $G(s) = G_0(s) + \delta G(s)$ $\qquad$ (1.3)

INPUT MULTIPLICATIVE PERTURBATION $\quad$ $G(s) = G_0(s) (I + R(s))$ $\qquad$ (1.4)

OUTPUT MULTIPLICATIVE PERTURBATION $\quad$ $G(s) = (I + L(s)) G_0(s)$ $\qquad$ (1.5)

The data given for Robust Control are :

- Nominal $G_0(s)$;
- Bounds on perturbation $\delta G(s), R(s), L(s)$,

$$\text{i.e. } \|\delta G(j\omega)\| \leq |g(j\omega)| \ , \quad \text{etc.} \qquad\qquad (1.6)$$

Frequency domain STRUCTURED uncertainty

In this case the uncertainty functions are of the form

$$\delta G(s) = g_1 G_1(s) + g_2 G_2(s) + \ldots + g_n G_n(s) \qquad\qquad (1.7)$$

and the data given for the design of a robust controller are:

- Nominal $G_0(s), G_1(s), \ldots\ldots G_n(s)$;
- upper and lower bounds on g_i .

In the theory presented here, only unstructured frequency domain models will be considered.

EXAMPLE 1.1

- Plant with frequency domain additive unstructured uncertainty

$$G(s) = G_0(s) + \delta G(s)$$

where

$$G_0(s) = \frac{1}{(1 - s)(s + 2)} \quad ; \quad \| \delta G(j\omega) \| < \frac{1}{\sqrt{\omega^2 + 9}}$$

- Plant with time domain structured uncertainty

$$\begin{cases} \dot{x}_1 = x_2 \\ \dot{x}_2 = g_1 x_1 + g_2 x_2 \end{cases}$$

with
$$10 < g_1 < 20 \quad ; \quad -.1 < g_2 < 0$$

$$A = \begin{pmatrix} 0 & 1 \\ g_1 & g_2 \end{pmatrix} = \begin{pmatrix} 0 & 1 \\ 0 & 0 \end{pmatrix} + g_1 \begin{pmatrix} 0 & 0 \\ 1 & 0 \end{pmatrix} + g_2 \begin{pmatrix} 0 & 0 \\ 0 & 1 \end{pmatrix}$$

□

Note that it is possible to design a compensator for a system with a structured uncertainty by using methods devoted to unstructured uncertainty systems, but the results will in general be too conservative. Structured uncertainty gives us much more information about the system than unstructured uncertainty. However, there are more complete theories for unstructured uncertainty in the frequency domain than for structured uncertainty. For this reason in the following only unstructured frequency domain uncertainty will be considered.

Readers interested in structured theories may consult :

- KHARITONOV Uncertain Polynomial Theory [2].
- DOYLE Structured Singular Value (SSV) Theory [18].

The problem of selecting nominal operating points and uncertainty
bounds for specific physical problems is nontrivial, but will not be
considered here.

The ROBUST CONTROL PROBLEM

> Given a nominal plant and perturbation bounds, find a __fixed__
> controller which yields a closed loop system with
> satisfactory performance for all "admissible" plant and
> disturbance signals.

Note: In ROBUST CONTROL the compensator is fixed and satisfies
performance requirements without any further adjustments. In contrast
ADAPTIVE CONTROL requires on line adjustment of the controller and
achieves satisfactory performance only asymptotically.

1.4 MATHEMATICAL PRELIMINARIES

In these lectures we limit our discussion to functions of a complex variable which are <u>rational</u> (i.e. ratios of polynomials).

Definition of Hardy spaces :

* <u>Hardy space</u>

   The space of all complex functions $F(s)$ of a complex variable s which are analytic in Re $s > 0$.

* <u>H^2- space</u>

   The space of Hardy functions for which the H^2 norm defined as

$$\|F(s)\|_2 \equiv \left(\int_{-\infty}^{+\infty} |F(j\ \omega)|^2 \ d\omega \right)^{1/2}$$

   is bounded, i.e.

$$\|F(s)\|_2 < \infty$$

Note : H^2 functions must be <u>strictly</u> <u>proper</u> and cannot have any poles on the $j\omega$-axis.

* <u>H^∞- space</u>

   The space of Hardy functions for which the H^∞ norm defined as

$$\|F(s)\|_\infty \equiv \sup_\omega \|F(j\omega)\|$$

   is bounded, i.e.

$$\|F(s)\|_\infty < \infty$$

NOTE: H^∞ functions must be <u>proper</u> and also cannot have any poles on the $j\omega$ axis. In addition from the maximum modulus theorem it follows that $|F(s)|$ is bounded for all s such that Re $s \geq 0$, since the maximum modulus of a function which is analytic in a region must be obtained on the contour of the region, unless the function is identically a constant. Finally a transfer function F(s) is BIBO stable if and only if it is an H^{∞} function.

Special H^{∞} functions

SCHUR
Complex H^{∞} function with the H^{∞} norm bounded by 1.

BOUNDED REAL
Real Schur function (Schur function with real coefficients).

STRICTLY BOUNDED REAL
Bounded real function with the H^{∞} norm strictly less than 1.

INNER FUNCTION
H^{∞} function F(s) with $|F(j\omega)| \equiv 1$ for all ω (all-pass function).

OUTER FUNCTION
H^{∞} function F(s) with all its zeros in Re s $\leq$ 0, including ∞ .

Fact: Every H^{∞} function can be written as the product of an inner function times an outer function.

This result is trivial in the scalar case. It will be extended in part II to MIMO systems.

Example 1.2

$$\frac{s(s-1)}{(s+3)^2(s+2)} = \left[\overset{\text{inner}}{\frac{s-1}{s+1}}\right]\left[\overset{\text{outer}}{\frac{(s+1)\ s}{(s+3)^2(s+2)}}\right]$$

Some other important H^∞ function are:

BLASCHKE PRODUCT

H^∞ functions of the form

$$B(s) = \prod \left(\frac{\alpha_i - s}{\overline{\alpha}_i + s} \right) \quad \text{where Re } \alpha_i > 0 .$$

Note: Blaschke products are inner functions.

UNIT

A Unit in H^∞ is an H^∞ function whose inverse is also H^∞.

Example 1.4

$$F(s) = \frac{1 - s}{1 + s} \quad \text{is not a unit in } H^\infty, \text{ since } 1/F(s) \text{ is not}$$

analytic in Re $s > 0$.

$$\frac{s + 1}{s + 2} \text{ is a unit in } H^\infty.$$

□

POSITIVE REAL FUNCTION

A function $Z(s)$ is a Positive Real (PR) function if:

      1) $Z(s)$ is analytic, for Re $s > 0$;

      2) Re $Z(s) \geq 0$, for Re $s > 0$;

      3) $Z(s)$ is real for s real.

Given a function $Z(s)$ define $S(s)$ as

$$S(s) = \frac{Z(s) - 1}{Z(s) + 1}$$

and the inverse relation

$$Z(s) = \frac{1 + S(s)}{1 - S(s)}$$

We have the following result:

Z(s) is a Positive Real function if and only if S(s) is a Bounded Real function.

STRICTLY POSITIVE REAL FUNCTION

A function Z(s) is Strictly Positive Real (SPR) if:

1) Z(s) is analytic for Re $s \geq 0$.
2) Re Z(s) > 0 for Re $s \geq 0$.
3) Z(s) is real for s real.

Example 1.3

	H^2	H^∞	BR	SCHUR	INNER	OUTER	Blasc prod.	PR	SPR
$\frac{2}{s + 1}$	*	*				*		*	*
$\frac{s}{s + 10}$		*	*	*		*		*	
$\frac{1 - s}{1 + s}$		*	*	*	*		*		
$\frac{1}{s^2 + 1}$									
$\frac{s + 1}{s + 2}$		*	*	*		*		*	*
$\frac{j}{s + 1}$	*	*			*	*		*	

* conditions are satisfied.

□

1.5 EXERCISES

1. Given a plant $G(s) = k/(s+a)$ where the uncertain parameters a and k vary between the limits $-2 \leq a \leq -1$ and $0.8 \leq k \leq 1.2$. Select the nominal plant to be $G_0(s) = 1/(s-1.5)$ and compute a bound on the uncertainty $|\delta G(j\omega)|$.

2. Given the second order function

$$F(s) = \frac{a\,s^2 + b\,s + c}{s^2 + d\,s + e}$$

Find necessary and sufficient conditions on the coefficient a, b, c, d, e for this function to be strictly positive real. Repeat for $F(s)$ to be an inner function.

3. Explain each of the blanks in the table of example 1.3 .

4. Show that if $F(s)$ and $G(s)$ are SPR so are $1/F(s)$ and $F(s) + G(s)$.

5. Show that if $F(s)$ and $G(s)$ are SBR so is $F(s)G(s)$.

CHAPTER 2

ROBUST STABILIZATION

2.1 NOMINAL INTERNAL STABILITY AND Q-PARAMETERIZATION

Consider the standard feedback configuration in fig.2.1

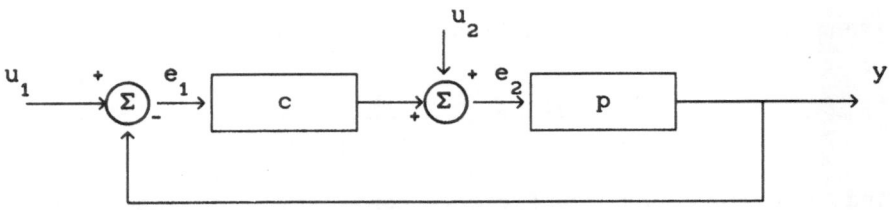

Fig. 2.1 General closed-loop system.

Denote now with h the transfer matrix between (u_1, u_2) and (e_1, e_2), then

$$h = \begin{pmatrix} h_{e_1,u_1} & h_{e_1,u_2} \\ h_{e_2,u_1} & h_{e_2,u_2} \end{pmatrix} = \begin{pmatrix} (1 + pc)^{-1} & -p(1 + cp)^{-1} \\ c(1 + pc)^{-1} & (1 + cp)^{-1} \end{pmatrix} \qquad (2.1)$$

The closed-loop system of fig.2.1 is said to be <u>internally stable</u> if all the transfer functions in matrix h are BIBO stable, i.e. H^{∞} functions. With this definition of stability we assure that any bounded signal injected at u_1 or at u_2 leads to a bounded responses at any other point of the system.

Introduce now the q function

$$q = \frac{c}{1 + pc} \ . \qquad (2.2)$$

The controller can be computed from q by the inverse relation

$$c = \frac{q}{1 - pq} \ . \qquad (2.3)$$

By using this relationship, matrix h becomes

$$
h = \begin{pmatrix} 1 - pq & -p(1 - pq) \\ q & 1 - pq \end{pmatrix}
\tag{2.4}
$$

In this way we can establish conditions on q for the internal stability of the system:

1) $q \in H^\infty$;
2) q must have zeros at poles of p in the RHP;
3) pq must interpolate to 1 at poles of p in the RHP.

Instead of looking for c in the design of a compensator we look for q. We develop next a parameterization of all q functions which satisfies the above conditions, following [27].

Assume for the moment that p(s) has no poles on $j\omega$ axes, let

$$
q(s) = B(s)\ \tilde{q}(s) \qquad \text{with } \tilde{q}(s) \in H^\infty
\tag{2.5}
$$

$$
\text{where } B(s) = \prod \left(\frac{\overline{\alpha}_i - s}{\overline{\alpha}_i + s} \right) \quad \text{(Blaschke product)}
\tag{2.6}
$$

and α_i are poles of p(s) in Re s > 0.

In order to simplify the treatment, we assume that all the poles are simple. With some modification, these results can be extended to the case of poles with a multiplicity greater than one.

With this choice conditions 1) and 2) are automatically satisfied.

Let us now consider

$$
\tilde{p}(s) = B(s)\ p(s)
\tag{2.7}
$$

then

$$
p(s)\ q(s) = \frac{\tilde{p}(s)}{B(s)}\ B(s)\ \tilde{q}(s) = \tilde{p}(s)\ \tilde{q}(s)
$$

condition 3) becomes

$$\tilde{q}(\alpha_1) = \frac{1}{\tilde{p}(\alpha_1)} = \beta_1 \qquad\qquad (2.8)$$

The problem of nominal internal stability design is reduced to the following interpolation problem :

Find a function $\tilde{q}(s) \in H^\infty$ such that $\tilde{q}(s)$ interpolates to β_1 at $s=\alpha_1$.

To compute the compensator

$$q(s) = B(s)\,\tilde{q}(s) \quad \text{and then} \quad c(s) = \frac{q(s)}{1 - p(s)q(s)} \ .$$

Example 2.1

Given the plant

$$p(s) = \frac{1}{(1 - s)\,(2 - s)}$$

we have

$$B(s) = \frac{(1 - s)\,(2 - s)}{(1 + s)\,(2 + s)}$$

$$\tilde{p}(s) = B(s)\,p(s) = \frac{1}{(s + 1)\,(s + 2)}$$

Then the interpolation conditions for $\tilde{q}(s)$ are

$$\tilde{q}(1) = \frac{1}{\tilde{p}(1)} = 6 \ ; \qquad \tilde{q}(2) = \frac{1}{\tilde{p}(2)} = 12.$$

To find a stabilizing compensator for $p(s)$ we only need to find an H^∞ function $\tilde{q}(s)$ which satisfies the above interpolation conditions, then apply the relation to compute $c(s)$. We discuss the solution of this interpolation problem next.

□

Solution to the H^∞ interpolation problem

The problem is, find a function $\tilde{q}(s) \in H^\infty$ such that

$$\tilde{q}(\alpha_i) = \beta_i.$$

Solution

Let

$$\tilde{q}(s) = \frac{n(s)}{d(s)} \qquad (2.9)$$

Pick d(s) as any Hurwitz polynomial.

(Hurwitz polynomial: a polynomial whose zeros all have negative real parts)

Find n(s) such that

$$n(\alpha_i) = \beta_i d(\alpha_i) = \gamma_i \qquad (2.10)$$

The problem is now reduced to polynomial interpolation.

A solution to this problem is given by the LAGRANGE INTERPOLATION FORMULA

$$n(s) = \gamma_1 \prod_{j \neq 1} \frac{(s - \alpha_j)}{(\alpha_1 - \alpha_j)} + \gamma_2 \prod_{j \neq 2} \frac{(s - \alpha_j)}{(\alpha_2 - \alpha_j)} + \ldots \qquad (2.11)$$

Example 2.2 (Lagrange interpolation)

Find a polynomial n(s) such that n(1)= .5 and n(2) = 4.

Solution

We have $\alpha_1 = 1$, $\alpha_2 = 2$; $\gamma_1 = .5$, $\gamma_2 = 4$;

by substituting in the Lagrange interpolation formula

$$n(s) = .5 \frac{(s - 2)}{(1 - 2)} + 4 \frac{(s - 1)}{(2 - 1)} = 3.5\,s - 3 \ .$$

□

Example 2.3

From example **2.1**, to find an H^∞ function $\tilde{q}(s)$ such that

$$\tilde{q}(1) = 6 \quad \text{and} \quad \tilde{q}(2) = 12 ;$$

we factor $\tilde{q}(s)$ as

$$\tilde{q}(s) = \frac{n(s)}{d(s)} ;$$

and we arbitrarily choose $d(s) = (s+1)^2$.

By this choice the problem is now that of finding a polynomial $n(s)$ such that

$$n(1) = \tilde{q}(1)d(1) = 24 ;$$

$$n(2) = \tilde{q}(2)d(2) = 108 .$$

From the Lagrange interpolation formula with

$$\alpha_1 = 1 , \quad \alpha_2 = 2 ; \quad \gamma_1 = 24 , \quad \gamma_2 = 108 ;$$

we have

$$n(s) = 24 \frac{(s-2)}{(1-2)} + 108 \frac{(s-1)}{(2-1)} = 84 s - 60$$

Then

$$\tilde{q}(s) = \frac{84 s - 60}{(s+1)^2} ;$$

$$q(s) = \tilde{q}(s) B(s) = \frac{(1-s)(2-s)(84s-60)}{(2+s)(s+1)^3} ;$$

The controller for example 2.1 is then given by

$$c(s) = \frac{q(s)}{1 - p(s)q(s)} = \frac{84 s - 60}{s^2 + 8 s + 31} .$$

□

Parameterization of all the solutions

All the solutions $\tilde{q}(s)$ satisfying the interpolation conditions can be expressed by

$$\tilde{q}(s) = \tilde{q}_A(s) + B(s)\, \tilde{q}_1(s) \qquad (2.12)$$

where $\tilde{q}_A(s)$ is a particular solution of our interpolation problem and $\tilde{q}_1(s)$ is an arbitrary H^∞ function.

Given a particular solution, computed, for example, with the Lagrange interpolation formula, by this formula we can express all the solutions to our problem. One can show that $\tilde{q}(s)$ given by (2.12) yields a q(s) which satisfies all the conditions for internal stability for any $\tilde{q}_1(s) \in H^\infty$.

Example 2.4 DISTURBANCE REJECTION [45]

Consider now the system shown in Fig.2.2

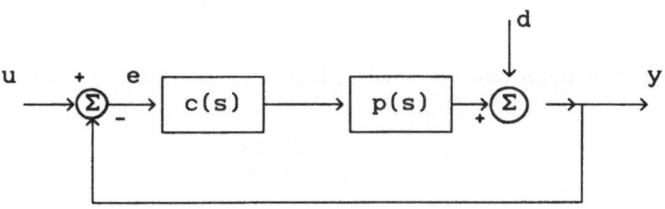

Fig. 2.2 Closed-loop system with disturbance input.

The transfer function between the disturbance and the output is

$$y/d = \frac{1}{1 + c(s)p(s)}$$

Substituting (2.3) in the above relation, we have

$$y/d = 1 - p(s)q(s) = 1 - p(s)\left(B(s)\, \tilde{q}_A(s) + B(s)^2\, \tilde{q}_1(s) \right) =$$

$$= T_1(s) + \tilde{q}_1(s) \, T_2(s) \; ;$$

where

$$T_1(s) = 1 - p(s)B(s)\tilde{q}_A(s) \; ;$$
$$T_2(s) = - p(s)B(s)^2;$$

are known and fixed H^∞ functions.

To design a compensator it will be sufficient to choose an H^∞ function $\tilde{q}_1(s)$ such that

$$e/d = T_1(s) + \tilde{q}_1(s) \, T_2(s) \qquad (2.13)$$

satisfies our performance requirements given for disturbance rejection. As discussed further in chapter 4, a possible performance measure for disturbance rejection is the H^∞ norm of the transfer function given in (2.13). The use of a free H^∞ function, such as $\tilde{q}_1(s)$ above, to parameterized all nominally stabilizing compensators for feedback design is referred to as Q-parameter design [43].

□

Example 2.5

Find all compensators which stabilize the nominal plant

$$p(s) = \frac{(s-2)(s+1)}{(s-1)(s-3)} \; ;$$

<u>Solution :</u>

STEP 1

Compute

$$B(s) = \frac{(s-1)(s-3)}{(s+1)(s+3)} \; ; \qquad \tilde{p}(s) = p(s) \, B(s) = \frac{(s-2)}{(s+3)} \; ;$$

STEP 2

Find an H^∞ function $\tilde{q}(s)$ such that

$$\tilde{q}(1) = \frac{1}{\tilde{p}(1)} = - 4 \; ; \qquad \tilde{q}(3) = \frac{1}{\tilde{p}(3)} = 6 \; ;$$

To solve the above interpolation problem we express $\tilde{q}(s)$ as

$$\tilde{q}(s) = \frac{n(s)}{s + 1} \; ;$$

Where $n(s)$ has to satisfy

$$n(1) = d(1)\,\tilde{q}(1) = -8 \; ; \qquad n(3) = d(3)\,\tilde{q}(3) = 24 \; ;$$

By the Lagrange interpolation formula we have

$$n(s) = -8 \; \frac{(s - 3)}{(1 - 3)} + 24 \; \frac{(s - 1)}{(3 - 1)} = 8(2s - 3);$$

so we have

$$\tilde{q}(s) = \frac{8(2s - 3)}{s + 1} \; ;$$

STEP 3

All solution $q(s)$ which yields to a stable compensator can then be parameterized as

$$q(s) = B(s)\,\tilde{q}(s) + B(s)^2\, q_1(s) \; ;$$

where $q_1(s)$ is an arbitrary H^∞ function.

Then the family of all compensator which stabilize $p(s)$ is

$$c(s) = \frac{q(s)}{1 - p(s)q(s)} = \frac{-(8(2s-3)(s+3) + (1-s)(3-s)q_1(s))}{(s+1)\,(\,15(s+3) + (s-2)\,q_1(s)\,)} \; .$$

□

2.2 STRONG STABILIZATION

A plant is said to be *strongly stabilizable* if it can be stabilized using a stable compensator [41]. Although the concept of strong stabilization is not directly a robust stability concept, it does relate to a number of robust stability concepts, most notably that of simultaneous stabilization which is discussed in the next section.

If the plant is stable we can always find a stable compensator. For unstable open-loop plants the conditions for the existence of a stable compensator are given by the following theorem :

Theorem 2.1 [41]

A dynamical plant $p(s)$ is strongly stabilizable if and only if the number of poles of $p(s)$ between every pair of real zeros, including infinity, of $p(s)$ in Re $s \geq 0$ is even.

This property is called the parity interlacing property (p.i.p.).

Example 2.6

$$p(s) = \frac{(s-1)(s-4)}{(s-2)(s-3)}$$

Strongly stabilizable

(Even number of poles between 1 and 4).

$$p(s) = \frac{(s-1)(s-4)}{(s+2)(s-3)}$$

Not strongly stabilizable

(Odd number of poles between 1 and 4).

$$p(s) = \frac{(s - 1)}{(s - 2)(s - 3)}$$

(pole-zero plot: zero o at 1, poles x at 2 and 3 on real axis)

Strongly
stabilizable

(Even number of poles between 1 and ∞).

$\square$

- **Sensitivity functions as the design parameter for strong stabilization [41].**

  Denote with $S(s)$ the sensitivity function, then

$$S(s) = \frac{1}{1 + c(s)p(s)} \quad \text{and} \quad c(s) = \frac{1 - S(s)}{p(s)S(s)} \qquad (2.14)$$

and matrix h in terms of sensitivity function becomes

$$h = \begin{pmatrix} S & -pS \\ \dfrac{1-S}{p} & S \end{pmatrix} \qquad (2.15)$$

The conditions on $S(s)$ for internal stability are (assume for the sake of simplicity that $p(s)$ has only simple poles in Re $s \geq 0$):

1) $S(s) \in H^{\infty}$;
2) poles of $p(s)$ in the RHP must be cancelled by zeros cf $S(s)$;
3) $S(s) = 1$ at zeros of $p(s)$ in the RHP.

As can be seen from $c(s)$ given in (2.14), for $c(s)$ to be stable we need to satisfy another condition:

4) zeros of $S(s)$ in the RHP must be <u>precisely</u> poles of $p(s)$ in the RHP.

If we now express $p(s)$ as

$$p(s) = \frac{n_p(s)}{d_p^{+}(s) \, d_p^{-}(s)} \qquad (2.16)$$

Where

$$d_p^+(s) = \text{a polynomial with all zeros in the RHP including the } j\omega \text{ axis.}$$

$$d_p^-(s) = \text{a polynomial with all zeros inside the LHP.}$$

In order to satisfy condition (1)-(4) , S(s) must be of the form

$$S(s) = \frac{\varepsilon \, d_p^+(s) \, h(s)}{g(s)} \qquad (2.17)$$

where $\varepsilon = \pm 1$, h(s) and g(s) are two strict Hurwitz polynomials satisfying

$$\delta(\, d_p^+ \,) + \delta(\, h \,) \le \delta(\, g \,) \qquad (2.18)$$

($\delta(\cdot)$ denotes the degree of polynomial).

Example 2.7

Find a stable compensator which stabilizes

$$p(s) = \frac{-\,0.164 \, (s + 0.2)(s - 0.32)}{s \, (s + 0.25)(s - 0.009)} \; ;$$

Solution :

STEP 1

From (10) we have

$$d_p^+(s) = s(s-0.009) \; ;$$

so from (2.11) S(s) can be chosen of the form

$$S(s) = \frac{d_p^+(s) \, (s + a)}{(s + 1)^3} \; ;$$

in this manner relation (2.12) is satisfied.

STEP 2

   At this point we compute the parameter 'a' so that condition 3
   is satisfied

$$S(\infty) = 1 ; \quad \text{and} \quad S(0.32) = 1 .$$

   It follows that

$$a = 22.791;$$

   will satisfy these conditions and we obtain finally

$$S(s) = \frac{s(s - 0.009)(s + 22.791)}{(s + 1)^3} .$$

STEP 3

   A stable stabilizing compensator for the plant p(s) is then
   given by

$$c(s) = \frac{1 - S(s)}{p(s)S(s)} = \frac{120.62 \, (s + 0.158) \, (s + 0.25)}{(s + 2) \, (s + 22.791)} .$$

□

– Strong stabilization via units in H^{∞} [39]

Fact: Every rational function can be expressed as a ratio of two
      H^{∞} functions [37].

Example 2.8

$$p(s) = \frac{(s-1)(s-2)}{(s-3)^2} = \frac{\dfrac{(s-1)(s-2)}{(s+1)^2}}{\dfrac{(s-3)^2}{(s+1)^2}}$$

□

Now express p(s) as

$$p(s) = \frac{N(s)}{D(s)} ; \qquad N(s), D(s) \in H^{\infty}$$

then consider

$$c(s) = \frac{U(s) - D(s)}{N(s)}$$

where U(s) is a UNIT in H^∞.

For c(s) to be stable U(s) must interpolate D(s) at zeros of N(s) in the RHP (zeros of p(s) in the RHP).

Note that with this choice the conditions for internal stability are automatically satisfied since all the elements of h

$$h = \begin{pmatrix} \dfrac{D}{U} & -\dfrac{N}{U} \\ \left(\dfrac{U-D}{N}\right)\dfrac{D}{U} & \dfrac{D}{U} \end{pmatrix}$$

are BIBO stable.

Proof : From the definition of unit $\frac{1}{U}$ is H^∞. The product of two H^∞ functions is also an H^∞ function. Finally by construction D and N are H^∞, and from the interpolation properties of U(s) the term $\left(\frac{U-D}{N}\right)$ is also H^∞.

Example 2.9

Find a stable compensator to stabilize

$$p(s) = \frac{(s - 1)}{s(s - .5)}$$

Solution :

The plant described by p(s) satisfy parity interlacing property, thus it is strongly stabilizable.

We factor p(s) as

$$p(s) = \frac{N(s)}{D(s)} \; ;$$

where we set

$$N(s) = \frac{s - 1}{(s + 1)^2} \; ; \qquad D(s) = \frac{s \, (s - .5)}{(s + 1)^2} \; ;$$

Now we need a Unit U(s) which interpolates to

$$U(1) = D(1) = \frac{1}{8} \; ;$$

$$U(\infty) = D(\infty) = 1 \; ;$$

a unit satisfying the above interpolation condition is

$$U(s) = \frac{s + 1}{s + 15}$$

Thus the transfer function of the compensator is

$$c(s) = \frac{U(s) - D(s)}{N(s)} = \frac{-(11.5 \, s + 1 \;)}{s + 15}$$

□

Example 2.10

Find a stable compensator to stabilize

$$p(s) = \frac{(s-1)(s-4)}{(s-2)(s-3)}$$

Solution :

Also the plant in this example satisfy parity interlacing property, thus it is strongly stabilizable.

First choose

$$N(s) = \frac{(s-1)(s-4)}{(s+1)^2} \; ;$$

$$D(s) = \frac{(s-2)(s-3)}{(s+1)^2} \; .$$

The equivalent interpolation problem is

Find a Unit in H^∞ which interpolates to

$$U(1) = D(1) = \frac{1}{2} \; ;$$

$$U(4) = D(4) = \frac{2}{25} \; ;$$

then compute the compensator with

$$c(s) = \frac{U(s) - D(s)}{N(s)} \; .$$

Finding a unit in H^∞ which interpolates to the above two points is nontrivial.

◻

Methods for finding the solution of this and other interpolation problems are developed in Chapter 3.

<u>FACT [37]</u>: If $U(\sigma_i)$ has the same sign for all σ_i, where σ_i are the zeros of $N(s)$ on the positive real axes, then a UNIT exists.

<u>FACT [37]</u>: If $p(s)$ satisfies p.i.p. then all $D(\sigma_i) > 0$ or all $D(\sigma_i) < 0$.

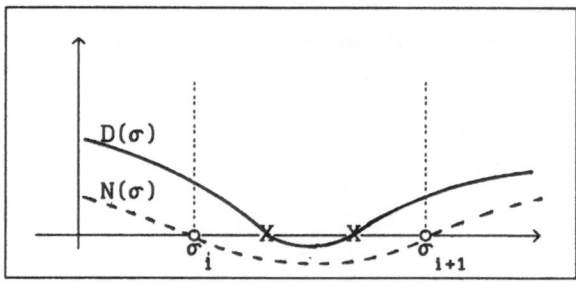

Fig. 2.3   Pole-zero interlacing.

From Fig. 2.3 we can see that if the zeros of $D(s)$ (poles of $p(s)$) between the zeros of $N(s)$ (zeros of $p(s)$) are even, p.i.p. is satisfied, then if $D(\sigma_i) > 0$ then it follows that $D(\sigma_{i+1}) > 0$.

- **Parameterization of all solutions**

<u>FACT [37]:</u>

A family of units which solve a given interpolation problem with only finite interpolation points can be parameterized in terms of an arbitrary strictly bounded real function

$$U(s) = \left(1 + B_z(s) \, V(s) \right) U_p(s) \qquad (2.19)$$

where

$B_z(s)$ is the Blaschke product of zeros of $N(s)$ in RHP;

$V(s)$ is any arbitrary SBR function ;

$U_p(s)$ is a particular solution satisfying the interpolation conditions.

Proof:

From $V(s)$ SBR we have

$$\left| B_z(s) \, V(s) \right| = \left| V(s) \right| < 1 \text{ in the RHP ;}$$

so $1 + B_z(s) \, V(s)$ has no zeros in the RHP , thus $1 + B_z(s) \, V(s)$ is a unit in H^∞. Since the product of two units is also a unit, it then follows that $\left(1 + B_z V \right) U_p$ is a unit. Finally since $B_z(\sigma_i)=0$, where σ_i are the interpolation points, it follows that $U(\sigma_i)=U_p(\sigma_i)$, independently of $V(s)$.

□

If $U(s)$ given by (2.13) is substituted back into the formula for the compensator

$$c(s) = \frac{U(s) - D(s)}{N(s)}$$

it is clear that strongly stabilizing compensators can parameterized in terms of an arbitrary strictly bounded real function $V(s)$.

Note: If $N(s)$ has a zero at infinity, then interpolation conditions are required at infinity and the term $B_z(s)$ must be appropriately modified. For example if $N(s)$ is of relative degree 1, then B_z can be replaced by $\dfrac{B_z}{s + 1}$, where B_z is the Blaschke product of all the finite zeros of $N(s)$ in the RHP.

2.3 SIMULTANEOUS STABILIZATION

SIMULTANEOUS STABILIZATION PROBLEM:

Find a fixed controller which simultaneously stabilizes the $\ell+1$ plants

$$G_0, G_1, \ldots, G_\ell$$

This problem has many important applications, for example, the design of a fixed controller which simultaneously stabilizes a set of linearized plants, the design of a fixed controller which preserves stability in the presence of actuator and sensor failures, etc.

Theorem [39]

$\ell + 1$ plants G_0, $G_1, \ldots, G_\ell$ can be stabilized by a fixed compensator, if and only if ℓ associated plants can be stabilized by a stable compensator.

Corollary (Two-plant case)

If G_0 is stable a compensator which stabilizes G_0 and G_1 exists if and only if there exists a stable compensator which stabilizes $G_1 - G_0$.

Proof of Corollary

If $G_0(s)$ is stable no interpolation conditions are required and the compensator which stabilizes $G_0(s)$ can be taken as, see (2.3),

$$c(s) = \frac{R(s)}{1 - G_0(s)R(s)} \quad ; \quad R(s) \in H^\infty . \tag{2.20}$$

Now the problem is to pick an H^∞ function $R(s)$ such that $c(s)$ also stabilizes $G_1(s)$.

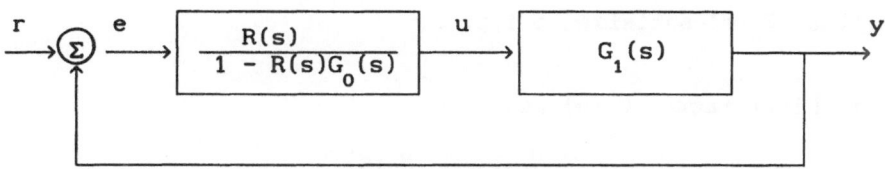

Fig. 2.4 Closed-loop system for plant $G_1(s)$.

From Fig. 2.4 we see that

$$y/r = \frac{\dfrac{R(s)}{1 - R(s)G_0(s)}\; G_1(s)}{1 + \dfrac{R(s)}{1 - R(s)G(s)}\; G_1(s)} = \frac{R(s)G_1(s)}{1 + R(s)\left[G_1(s) - G_0(s)\right]}$$

To stabilize two plants simultaneously is thus equivalent to finding a __stable__ $R(s)$ which stabilizes the difference of the two plants transfer function $G_1(s) - G_0(s)$, the compensator is then given by

$$c(s) = \frac{R(s)}{1 - G_0(s)R(s)}$$

Note that, for a solution to exist, $G_0(s) - G_1(s)$ must satisfy the p.i.p. .

□

Example 2.11
Find a compensator which stabilizes the two plants $G_0(s)$ and $G_1(s)$

$$G_0(s) = \frac{1}{s + 1}\quad ;\qquad G_1(s) = G_0(s) + \frac{(s-1)(s-3)}{(s-2)^2}\;.$$

We have to compute a compensator which strongly stabilizes the difference plant

$$G_D(s) = G_1(s) - G_0(s) = \frac{(s-1)(s-3)}{(s-2)^2}$$

Note that $G_D(s)$ satisfies p.i.p. .

<u>Step 1</u> First factor $G_D(s)$ as

$$G_D(s) = \frac{N_A(s)}{D_A(s)}$$

with

$$N_A(s) = \frac{(s-1)(s-3)}{(s+1)^2} \quad \text{and} \quad D_A(s) = \frac{(s-2)^2}{(s+1)^2}$$

<u>Step 2</u> Then find a unit $U(s)$ in H^∞ such that

$$U(1) = D_A(1) = \frac{1}{4} \; ;$$

$$U(3) = D_A(3) = \frac{1}{16} \; .$$

<u>Step 3</u> Compute $R(s)$

$$R(s) = \frac{U(s) - D_A(s)}{N_A(s)}$$

The compensator $c(s)$ is then given by

$$c(s) = \frac{R(s)}{1 - G_0(s)R(s)}$$

As previously noted, finding a unit which interpolates the above two points is nontrivial. Indeed it is easy to show that no first order unit can interpolate these two point by exploring units of the form $U(s)=(as+b)/(s+c)$ where a,b,c are constrained to be positive.

□

A general algorithm for interpolating with units will be given in chapter 3.

2.4 ROBUST STABILIZATION : ADDITIVE AND MULTIPLICATIVE UNSTRUCTURED PLANT PERTURBATIONS

PLANT WITH ADDITIVE UNSTRUCTURED PERTURBATIONS

Definition [26]

A transfer function $p(s)$ is said to be in the class $C(p_0(s), r(s))$ if

i) $p(s)$ has the same number of unstable poles as $p_0(s)$;

ii) $|p(j\omega) - p_0(j\omega)| \leq |r(j\omega)|$, $|r(j\omega)| > 0 \quad \forall \omega$.

$\square$

Our problem is to find a compensator which stabilizes all the plants in a class $C(p_0(s), r(s))$.

The Robust Stability Condition

Given a compensator $c(s)$ which stabilizes $p_0(s)$ we can establish the conditions for $c(s)$ to be a robust stabilizer for all the plants in the class $C(p_0(s), r(s))$.

Denote with

$$p(s) = p_0(s) + \delta p(s) \quad ; \quad |\delta p(j\omega)| \leq |r(j\omega)| \quad \forall \omega$$

a generic plant in the class $C(p_0(s), r(s))$ and we assume further that $r(j\omega)$ has no finite zeros for real ω.

From the hypothesis that $c(s)$ stabilizes $p_0(s)$ we have

$$p_0(j\omega)c(j\omega) + 1 \neq 0 \quad \forall \omega$$

and $p_0(j\omega)c(j\omega)$ has the correct encirclements of the -1 point to guarantee, from Nyquist's stability criterion, that the nominal closed-loop system is stable. If $p_0(s)$ and the perturbed plant $p(s)$ have the same number of unstable poles then

$$1 + p(j\omega)c(j\omega)$$

can be written

$$(1 + p_0(j\omega)c(j\omega))(1 + (1 + p_0(j\omega)c(j\omega))^{-1} c(j\omega) \delta p(j\omega))$$

and from the small gain theorem [21] one is guaranteed that the number of encirclements of the -1 point will not change with the perturbed plant p(s) as long as

$$\sup_{\omega} | (1 + p_0(j\omega)c(j\omega))^{-1} c(j\omega) \delta p(j\omega) | < 1$$

A sufficient condition for robust stability is then

$$\| (1 + p_0(s)c(s))^{-1} c(s) r(s) \|_{\infty} < 1 \qquad (2.21)$$

Actually if arbitrary complex values of δp are permitted at each frequency the condition (2.21) is also necessary for robust stability [37].

If we introduce the q parameter

$$q(s) = \frac{c(s)}{1 + p_0(s)c(s)}$$

then the robust stability condition on q(s) is

$$\| q(s)r(s) \|_{\infty} < 1 \qquad (2.22)$$

Now recall the results of Sec. 2.1 on q(s) for c(s) to guarantee internal stability, i.e.

$$q(s) = B(s) \tilde{q}(s)$$

where B(s) is the Blaschke product of poles of $p_0(s)$ in RHP (assume that $p_0(s)$ has no poles on the jω axes)

$$\tilde{p}_0(s) = B(s)p_0(s)$$

$$p_0(s)q(s) = \tilde{p}_0(s)\tilde{q}(s)$$

and $\tilde{q}(s)$ has to satisfy the interpolation conditions

$$\tilde{q}(\alpha_i) = \frac{1}{\tilde{P}_0(\alpha_i)}$$

Now denote with $r_m(s)$ a minimum phase H^∞ function (outer function with no zeros on the $j\omega$-axis) such that

$$|r_m(j\omega)| = |r(j\omega)|.$$

The robust stability condition can then be written

$$\|\tilde{q}(s) \; r_m(s)\|_\infty < 1$$

where $r_m(s)$ is a minimum phase function.

Now introduce the function

$$u(s) = \tilde{q}(s) \; r_m(s) \; .$$

The robust stability condition becomes

$$\|u(s)\|_\infty < 1 \; . \tag{2.23}$$

Since $r_m(s)$ and $\tilde{q}(s)$ are H^∞, the above relation implies that $u(s)$ must be an SBR function.

The interpolation conditions on $u(s)$ are

$$u(\alpha_i) = \tilde{q}(\alpha_i) \; r_m(\alpha_i) = \frac{r_m(\alpha_i)}{\tilde{P}_0(\alpha_i)} = \beta_i \; . \tag{2.24}$$

The robust stability problem is reduced to the following equivalent interpolation problem:

Find an SBR function $u(s)$ which interpolates to given points in the RHP. In the mathematical literature is known as the Nevanlinna-Pick interpolation problem [11].

The robustly stabilizing compensator c(s) it is then computed from the following steps :

Step 1 Compute $r_m(s)$, $\tilde{p}_0(s)$ and B(s) ;

Step 2 Find an SBR function u(s) such that $u(\alpha_i) = \beta_i = \dfrac{r_m(\alpha_i)}{\tilde{p}_0(\alpha_i)}$;

(Solve the Nevanlinna-Pick problem)

where α_i are the RHP poles of $p_0(s)$.

Step 3 Compute $q(s) = \dfrac{B(s)u(s)}{r_m(s)}$;

Step 4 Compute $c(s) = \dfrac{q(s)}{1 - p_0(s)q(s)}$.

With respect to step 3), for q(s) to be an H^∞ function we must assure that

$$\frac{u(s)}{r_m(s)} \in H^\infty .$$

Since u(s) and B(s) are H^∞ functions the only function which can cause a problem is $1/r_m(s)$. Two cases exists.

1) $r_m(s)$ is a unit (Exactly proper, minimum phase function).

From the definition of unit $\dfrac{1}{r_m(s)} \in H^\infty$ and

q(s) is a product of three H^∞ functions hence also an H^∞ function.

2) $\lim\limits_{s \to \infty} r_m(s) = 0$ but $\lim\limits_{s \to \infty} s\, r_m(s) \neq 0$ (The relative degree of $r_m(s)$ is 1).

In this second case another interpolation condition on u(s) is needed

$$u(\infty) = 0 .$$

If $r_m(s)$ has a relative degree greater than one the interpolation problem at ∞ is more complex, but can in principle, be solved.

Plants with one pole at the origin

If $p_0(s)$ has a pole at s=0 it is possible to extend the method described above. If it is assumed that the pole is preserved under perturbations then the uncertainty bound r(s) must also have a pole at s=0.

So it is possible to write

$$r(s) = \frac{r'_m(s)}{s} \; ;$$

where $r'_m(s)$ is a minimum phase H^∞ function.

If we now pose

$$\tilde{p}(s) = s\, B(s)\, p_0(s) \; ;$$

and

$$\tilde{q}(s) = \frac{q(s)}{s\, B(s)} \; ;$$

the robust stability condition becomes

$$\|q(s)r(s)\|_\infty = \|\tilde{q}(s)r'_m(s)\|_\infty < 1 \; .$$

If we pose

$$u(s) = \tilde{q}(s)\, r'_m(s)$$

the problem is now that of finding an SBR function u(s) with the interpolation conditions

$$u(\alpha_i) = \beta_i \; ; \qquad \text{where} \qquad \beta_i = \frac{r'_m(\alpha_i)}{\tilde{p}_0(\alpha_i)} \; ;$$

and

$$u(0) = \beta_0 \; ; \qquad \text{where} \qquad \beta_0 = \frac{r'_m(0)}{\tilde{p}_0(0)} \; .$$

With these interpolation conditions we guarantee that

$$p_0(s)q(s) = \tilde{p}_0(s)\tilde{q}(s)$$

is equal to one at the unstable poles of $p_0(s)$, so the compensator $c(s)$ is guaranteed to stabilize $p_0(s)$.

Plants without poles in the Right Half Plane

If $p_0(s)$ is a stable plant, robust stability conditions are simpler. There are no interpolation conditions and the only condition to be satisfied is

$$\| \, q(s) \, r_m(s) \|_\infty < 1$$

If we pose

$$u(s) = q(s)r_m(s)$$

to design a robust controller it is only necessary to choose $u(s)$ as any SBR function. Then $q(s)$ is given by

$$q(s) = \frac{u(s)}{r_m(s)}$$

and the compensator by

$$c(s) = \frac{q(s)}{1 - p_0(s)q(s)}$$

If the relative degree of $r_m(s)$ is greater than zero, then $u(s)$ does have to meet interpolation conditions at ∞.

PLANT WITH MULTIPLICATIVE UNSTRUCTURED PERTURBATIONS

Definition [37]

A transfer function $p(s)$ is said to be in the class $M(p_0(s), r(s))$
if :

1) $p(s)$ has the same number of unstable poles as $p_0(s)$;

2) $p(s) = (1 + M(s)) p_0(s)$ with $|M(j\omega)| < |r(j\omega)|$ $\forall\omega$.

where $r(s)$ is constrained to be a rational function with no finite
unstable poles or zeros.

The Robust stability condition

A compensator $c(s)$ which stabilizes $p_0(s)$ is a robust compensator for
all the plants in the class $M(p_0(s), r(s))$ if

$$\| p_0(s)c(s) (1 + p_0(s)c(s))^{-1} r(s) \|_\infty < 1 \qquad (2.25)$$

Proof:

From the hypothesis that $c(s)$ stabilizes $p_0(s)$ we have

$$p_0(j\omega)c(j\omega) + 1 \neq 0 \qquad \forall\omega .$$

For $c(s)$ to be a robust compensator for all the plants in the class
$M(p_0(s), r(s))$ it is necessary that

$$|1 + p(j\omega)c(j\omega)| \neq 0 \qquad \forall\ \omega$$

By substituting

$$p(s) = (1 + M(s)) p_0(s)$$

we have

$$|1 + p(j\omega)c(j\omega)| =$$

$$= |(1+p_0(j\omega)c(j\omega))(1 + p_0(j\omega)c(j\omega)(1+p_0(j\omega)c(j\omega))^{-1} M(j\omega)|$$

and considering that

$$|M(j\omega)| < |r(j\omega)| \quad \forall \omega$$

we have from the small gain theorem that (2.18) is sufficient to guarantee robust stabilization. As in the case of additive perturbations (2.18) is also a necessary condition for robust stabilization if arbitrary complex perturbations M(s) are permitted.

Equivalent interpolation conditions

A robust compensator c(s) has to satisfy

$$\| \, p_0(s)c(s) \, (\, 1 + p_0(s)c(s) \,)^{-1} \, r(s) \, \|_\infty < 1 \; .$$

This condition expressed in term of the function q(s) becomes

$$\| \, p_0(s) \, q(s) \, r(s) \, \|_\infty < 1 \; . \tag{2.26}$$

Now denote with $B_z(s)$ the Blaschke product of the zeros of $p_0(s)$ inside the RHP, the above condition is then equivalent to

$$\left\| \frac{p_0(s) \, q(s) \, r(s)}{B_z(s)} \right\|_\infty < 1 \tag{2.27}$$

since $|B_z(j\omega)| \equiv 1$.
Now let

$$u(s) = \frac{p_0(s) \, q(s) \, r(s)}{B_z(s)} \; . \tag{2.28}$$

Note that the zeros of $B_z(s)$ in the RHP are cancelled by the zeros of $p_0(s)$ so that if q(s) and r(s) are H^∞ functions, u(s) will also be an H^∞ function.

To find a compensator which stabilizes all plants in the class
$M(p_0(s), r(s))$ it is then necessary to find an SBR function $u(s)$
which satisfies the interpolation conditions

$$u(\alpha_i) = \frac{r(\alpha_i)}{B_z(\alpha_i)} \qquad (2.29)$$

where α_i are the poles of $p_0(s)$ inside the RHP, since at the poles
of $p_0(s)$ in the RHP $p_0(s)q(s) = 1$.
Then one may compute $q(s)$ from the relation

$$q(s) = \frac{B_z(s)u(s)}{p_0(s)r(s)} \qquad (2.30)$$

Note that $B_z(s)$ in (2.30) guarantees that $q(s) \in H^\infty$.
With these choices the interpolation conditions on $q(s)$, for $c(s)$ to
be a stabilizing compensator for $p_0(s)$, are satisfied.

Note 1 : If the relative degree of the product $p_0(s)r(s)$ is greater
than zero, interpolation conditions will be required for $u(s)$
at ∞, as in the additive perturbation case.

Note 2 : The function $r(s)$ need not necessarily be proper, as long as
the product $p_0(s)r(s)$ is proper. Thus for example, $r(s)$ may be
a polynomial.

2.5 EXERCISES

1. Given the plant

$$P_0(s) = \frac{s - 2}{s - 12}$$

with multiplicative uncertainty

$$M(s) = \frac{1}{3} \left(\frac{s + 1}{s + 2} \right) \frac{(s + 6)^2}{(s^2 + 2s + 37)}$$

Find a compensator which is robustly stabilizing.

2. Find the equivalent interpolation problem corresponding to the simultaneous stabilization of the two plants

$$G_0(s) = \frac{1}{s + 1} \quad , \quad G_1(s) = \frac{s^3 - s^2 + 172s - 168}{(s + 1)(s^2 + 173s - 170)} \quad .$$

3. Find the equivalent interpolation problem corresponding to the robust stabilization of the nominal plant

$$P_0(s) = \frac{1}{(s - 1)(s - 2)(s - 3)}$$

given the additive uncertainty bound

$$r(s) = \frac{s + 0.1}{10s + 1} \quad .$$

CHAPTER 3

NEVANLINNA-PICK INTERPOLATION THEORY

3.1 INTERPOLATION WITH SCHUR AND BOUNDED REAL FUNCTIONS [11]

As was noted in the previous chapter, the problem of robust stabilization of an unstable system can be reduced to the problem of finding an SBR function u(s) which interpolates to

$$u(\alpha_i) = \beta_i \quad , \quad i=1,\ldots,\ell \; ; \; Re \; \alpha_i > 0 \; , \; |\beta_i| < 1 \; .$$

In this section we first describe the so-called Nevanlinna-Pick algorithm [11] for solving this problem. This algorithm in general gives us a Schur function which satisfies our interpolation conditions but not a Strictly Bounded Real function, because coefficients may be complex. Only when all α_i are real do we obtain with this algorithm an SBR function. However , as we shall remark later, an SBR function wich satisfies the same interpolation constraints can be computed from a Schur function by a simple formula. In order to simplify the formulation we assume that all points α_i are distinct and that there are no points α_i on the $j\omega$ axes. This hyphothesis can be removed but the resulting algorithm is significantly complicated.

Note that in much of the mathematical literature, for example [11], the Schur function in question is defined to be analytic inside the <u>unit circle</u> rather than the right-half plane, and the interpolation points are points inside the unit circle. It is a simple matter to map the RHP into the interior of the unit circle, e. g. via the transformation

$$z = \frac{s - 1}{s + 1} \; ,$$

so that the interpolation theory can be developed in either domains. Since we have defined our stability conditions for continuous-time systems, it will be more convenient to deal with the interpolation theory in the s-domain directly.

We develop next an algorithm for interpolation of points in the RHP with Schur functions and ultimately SBR functions.

Consider the mapping between two functions $u_1(s)$ and $u_2(s)$

$$u_1(s) = \frac{\rho_1 + u_2(s)\left(\dfrac{s - \alpha_1}{s + \bar{\alpha}_1}\right)}{1 + \bar{\rho}_1\left(\dfrac{s - \alpha_1}{s + \bar{\alpha}_1}\right)u_2(s)} = T_{\rho_1}^{-1}\left[u_2(s)\right] \quad ; \quad |\rho_1| < 1 \quad (3.1)$$

The following properties can easily be verified

<u>FACT 3.1</u> If $u_2(s)$ is a Schur function then $u_1(s)$ is also a Schur function, and $\quad u_1(\alpha_1) = \rho_1$, independently of $u_2(s)$.

For the inverse relation

$$u_2(s) = \frac{u_1(s) - \rho_1}{1 - \bar{\rho}_1 u_1(s)}\left(\frac{s + \bar{\alpha}_1}{s - \alpha_1}\right) = T_{\rho_1}\left[u_1(s)\right] \qquad (3.2)$$

we have

<u>FACT 3.2</u> If $u_1(s)$ is a Schur function and $u_1(\alpha_1)=\rho_1$ then $u_2(s)$ is a Schur function.

(Note that if α_1 is real the above mappings are from SBR functions to SBR functions).

Proofs :

FACT 3.1

$u_1(s)$ is a Schur function if and only if

a) $u_1(s) \in H^{\infty}$;

and

b) $1 - \bar{u}_1(j\omega) u_1(j\omega) \geq 0$;

Let

$$a = \rho_1; \qquad b(s) = u_2(s) \left(\frac{s - \alpha_1}{s + \bar{\alpha}_1} \right) \; ;$$

then

$$u_1(s) = \frac{a + b(s)}{1 + \bar{a}\, b(s)} \; ; \qquad\qquad (3.3)$$

from the fact that $u_2(s)$ is Schur, $|\bar{a}| < 1$, and the maximum modulus theorem it follows that

$$1 + \bar{a}\, b(s) \neq 0 \text{ for Re } s \geq 0$$

so that $u_1(s) \in H^\infty$.

Next, for $s = j\omega$, $\quad 1 - \bar{u}(j\omega)\, u(j\omega)$ can be written

$$1 - \bar{u}_1(j\omega)\, u_1(j\omega) = \frac{\left[1 - |a|^2 \right]\left[1 - |b(j\omega)|^2 \right]}{|\, 1 + \bar{a}\, b(j\omega)\, |^2} \; ;$$

which is nonnegative since by assumption $|a| < 1$ and $|b(j\omega)| \leq 1$.

◻

FACT 3.2

Similarly $u_2(s)$ is Schur if and only if:

a) $u_2(s) \in H^\infty$.

and

b) $1 - \bar{u}_2(j\omega)\, u_2(j\omega) \geq 0$.

Note that $u_2(s)$ given in (3.2) can be viewed as the product of two functions one of which is exactly of the form (3.3), i.e. the function

$$\frac{u_1(s) - \rho_1}{1 - \bar{\rho}_1\, u_1(s)}$$

and the other is the function

$$\left(\frac{s + \bar{\alpha}_1}{s - \alpha_1} \right) .$$

(3.4)

From the fact that

$$\left| \frac{j\omega + \bar{\alpha}_1}{j\omega - \alpha_1} \right| \equiv 1$$

it follows from exactly the same arguments as in the proof of fact 3.1 that $1 - \bar{u}(j\omega) u(j\omega) \geq 0$. In addition while it appears that the term (3.4) introduces a pole in the R.H.P., in reality the pole is cancelled by the interpolation condition $u_1(\alpha_1) = \rho_1$, thus $u_2(s)$ is analytic in the R.H.P. as required.

□

By this mapping we can reduce the problem of finding a Schur function interpolating ℓ points to the problem of finding a Schur function interpolating $\ell-1$ points. In this manner by repeating this mapping iteratively we can arrive at only one interpolation point.

Remember that our problem was that of finding a Strictly Bounded Real function u(s) such that $u(\alpha_1) = \beta_1$, $u(\alpha_2) = \beta_2$, ... $u(\alpha_\ell) = \beta_\ell$.
First we compute a Schur function which satisfies these constraints, then we extend the results to SBR functions. (If α_i are real we directly obtain an SBR function).
If now we pose

$$u_1(s) = u(s), \quad \rho_1 = \beta_1;$$

then by using the relation (3.2) we can compute

$$\rho_2 = u_2(\alpha_2) = T_{\rho_1}(u_1(\alpha_2)) = T_{\rho_1}(\beta_2) \; ;$$

$$\rho_{2,3} = u_2(\alpha_3) = T_{\rho_1}(u_1(\alpha_3)) = T_{\rho_1}(\beta_3) \; ;$$

$$\ldots \ldots \ldots \ldots \ldots \ldots \ldots \ldots \ldots \ldots$$

$$\rho_{2,\ell} = u_2(\alpha_\ell) = T_{\rho_1}(u_1(\alpha_\ell)) = T_{\rho_1}(\beta_\ell) \; ;$$

The problem is reduced to one of finding a Schur function $u_2(s)$ which interpolates to $\ell-1$ points

$$u_2(\alpha_2) = \rho_2 \; ; \quad u_2(\alpha_3) = \rho_{2,3} \; ; \quad \ldots \ldots \quad u_2(\alpha_\ell) = \rho_{2,\ell} \; ;$$

Then $u_1(s)$ is given from (3.1).

If we now generalize this procedure we arrive at the following algorithm :

Step 1) Compute all the elements ρ_i and $\rho_{i,j}$ of the so called <u>Fenyves</u> <u>array</u>

α_1	α_2	α_3	$\ldots \ldots$	α_ℓ	
ρ_1	$\rho_{1,2}$	$\rho_{1,3}$	$\cdots\cdots$	$\rho_{1,\ell}$	$u_1(s)$
	ρ_2	$\rho_{2,3}$	$\cdots\cdots$	$\rho_{2,\ell}$	$u_2(s)$
		$\ldots\ldots\ldots\ldots$			$\ldots\ldots$
				ρ_ℓ	$u_\ell(s)$

where every row is computed by using a generalization of the relation (3.2) :

For the first row

$$\rho_1 = \beta_1 \qquad \text{and} \quad \rho_{1,j} = \beta_j \; ;$$

for the other rows

$$\rho_i = T_{\rho_{i-1,1}}(\rho_{i-1,1}) = \frac{\rho_{i-1,1} - \rho_{i-1}}{1 - \bar{\rho}_{i-1}\,\rho_{i-1,1}} \cdot \frac{\alpha_i + \bar{\alpha}_{i-1}}{\alpha_i - \alpha_{i-1}} \quad ; \quad 1 < i \leq \ell$$

and

$$\rho_{i,j} = T_{\rho_{i-1}}(\rho_{i-1,j}) = \frac{\rho_{i-1,j} - \rho_{i-1}}{1 - \bar{\rho}_{i-1}\,\rho_{i-1,j}} \cdot \frac{\alpha_j + \bar{\alpha}_{i-1}}{\alpha_j - \alpha_{i-1}} \quad ; \quad 1 < i < j \leq \ell \; .$$

Step 2) Compute a Schur function $u_\ell(s)$ which interpolates to

$$u_\ell(\alpha_\ell) = \rho_\ell$$

Step 3) Compute iteratively from the relation

$$u_{i-1}(s) = T_{\rho_{i-1}}^{-1}(u_i(s)) = \frac{\rho_{i-1} + u_i(s)\left(\dfrac{s - \alpha_{i-1}}{s + \bar{\alpha}_{i-1}}\right)}{1 + \bar{\rho}_{i-1}\left(\dfrac{s - \alpha_{i-1}}{s + \bar{\alpha}_{i-1}}\right)u_i(s)}$$

all the functions

$$u_{\ell-1}(s), \; u_{\ell-2}(s) \; \ldots\ldots\ldots \; u_2(s) \; , \quad u_1(s) \; .$$

To solve step 2) we have two possibilities :
a) Choose

$$u_\ell(s) = \rho_\ell$$

In this manner we find only one solution.

b) Choose a parameterization of $u_\ell(s)$ in terms of an arbitrary Schur function $u_{\ell+1}(s)$

$$u_\ell(s) = T^{-1}_{\rho_\ell}(u_\ell(s)) = \frac{\rho_\ell + u_{\ell+1}(s)\left(\dfrac{s - \alpha_\ell}{s + \bar{\alpha}_\ell}\right)}{1 + \bar{\rho}_\ell\left(\dfrac{s - \alpha_\ell}{s + \bar{\alpha}_\ell}\right)u_{\ell+1}(s)} \ .$$

With this choice we can find the family of all the Schur functions which satisfy our interpolation condition parameterized in terms of an arbitrary Schur function. Note that this second option comprises the first if we choose $u_{\ell+1}(s) \equiv 0$, 0 being a Schur function.
Some conditions to test whether a solution exists are given from the theorem

Theorem 3.1 [11]
The Nevanlinna Pick problem admits solution if and only if the Pick matrix P, whose elements p_{ij} are given by the relation

$$p_{ij} = \frac{1 - \beta_i\bar{\beta}_j}{\alpha_i + \bar{\alpha}_j} \tag{3.5}$$

is non-negative definite.

If we require a solution strictly bounded by one in modulo then a solution exists if and only if the Pick matrix P is positive definite.

An equivalent result is the following.

Theorem 3.2 [40]
The Nevanlinna-Pick problem admits a solution if and only if all the elements of the Fenyves arrays are in modulo less than one, i.e.

$$|\rho_i| < 1 \qquad \text{and} \qquad |\rho_{i,j}| < 1 \ . \tag{3.6}$$

Note that as a consequence of the Maximum Modulus Theorem if we have $|\rho_k| = 1$ for some k then a solution $u_k(s) \equiv \rho_k$ only exists if $\rho_{k,j} = \rho_k$. In this case we have

$$|u_{k-1}(j\omega)| \equiv |u_{k-2}(j\omega)| \equiv \ldots \equiv |u_1(j\omega)| \equiv 1 \quad \text{all } \omega.$$

(Remember that $u_k(s)$ is a Schur function then $|u_k(j\omega)| \leq 1$. The maximum modulus theorem assures that because $u_k(s)$ is analytic in the RHP, it should assume its maximum values on the $j\omega$ axes; otherwise it is the constant function).

<u>Dealing with the points s=0 and s=∞</u>

In this case the Nevanlinna-Pick algorithm may be modified by extending the Fenyves array as follows

α_1	α_2	α_3	$\ldots$	α_ℓ	0	∞	
ρ_1	$\rho_{1,2}$	$\rho_{1,3}$	$\ldots$	$\rho_{1,\ell}$	$\rho_{1,\ell+1}$	$\rho_{1,\ell+2}$	$u_1(s)$
	ρ_2	$\rho_{2,3}$	$\ldots$	$\rho_{2,\ell}$	$\rho_{2,\ell+1}$	$\rho_{2,\ell+2}$	$u_2(s)$
		$\ldots\ldots\ldots\ldots\ldots$			$\ldots\ldots\ldots\ldots\ldots$		$\ldots\ldots$
				ρ_ℓ	$\rho_{\ell,\ell+1}$	$\rho_{\ell,\ell+2}$	$u_\ell(s)$
					$\rho_{\ell+1}$	$\rho_{\ell+1,\ell+2}$	$u_{\ell+1}(s)$

Then we can choose $u_{\ell+1}(s)$ as a Schur function which interpolates to

$$u_{\ell+1}(0) = \rho_{\ell+1} \quad \text{and} \quad u_{\ell+1}(\infty) = \rho_{\ell+1,\ell+2} \; ;$$

for example

$$u_{\ell+1}(s) = \frac{\rho_{\ell+1,\ell+2} \, s + \rho_{\ell+1}}{s + 1} .$$

Extension to BR and SBR functions

If some points α_i are complex then the above Nevanlinna-Pick algorithm gives a Schur function $u_s(s)$ which may have complex coefficients. A Bounded Real function $u_{BR}(s)$ with the same interpolation condition can be computed by use of the transformation

$$u_{BR}(s) = \frac{1}{2}\left(u_s(s) + \overline{u_s(\bar{s})} \right) ; \qquad (3.7)$$

However, this formula can result in a high-order interpolating function. A modified version of the Nevanlinna-Pick algorithm which directly gives a Bounded Real function of an order generally lower than the above formula is reported in [14].

3.2 INTERPOLATION WITH POSITIVE REAL FUNCTIONS AND UNITS IN H^∞

In sections 2.2 and 2.3 it was shown how strong and simultaneous stabilization problems can be reduced to interpolation problems with units in H^∞. In this section a modification of the Nevanlinna-Pick algorithm suitable for interpolation with positive real functions and units in H^∞ is presented.

This algorithm is based on the following two facts, here stated without proof.

Fact 3.3

If $Z(s)$ is a Strictly Positive Real (SPR) function and Exactly Proper (EP) (relative degree zero) (EP/SPR), then $Z(s)$ is a Unit in H^∞ as well as $\left[Z(s)\right]^m$.

Examples 3.1

$Z_1(s) = \dfrac{1}{s}$ is a Positive Real function but it is not Exactly proper.

$Z_2(s) = s$ is a Positive Real function but it is not Exactly proper.

$Z_3(s) = \dfrac{1}{s + 1}$ SPR but not EP.

$Z_4(s) = \dfrac{s + 1}{s + 2}$ SPR and EP : it is a Unit in H^∞.

□

Fact 3.4

A PR function $Z(s)$ interpolates to $Z(\alpha_i) = \beta_i$; $i = 1, \ldots, \ell$ if and only if

the BR function

$$S(s) = \frac{Z(s) - 1}{Z(s) + 1}$$

interpolates to

$$S(\alpha_i) = \frac{\beta_i - 1}{\beta_i + 1}$$

In this way a problem of interpolation with a Positive real function is reduced to a problem of interpolation with a BR function.

An algorithm for interpolation with Units in H^∞ can then be stated as follows.

Problem :

Find a unit $u(s)$ in H^∞ such that $u(\alpha_i) = \beta_i$; $i = 1, \ldots, \ell$.

Algorithm :

Step 1) Find an integer m such that the points

$$\sigma_i = \frac{\beta_i^{\frac{1}{m}} - 1}{\beta_i^{\frac{1}{m}} + 1}$$

can be interpolated with an SBR function. (For example with the Nevanlinna-Pick algorithm)

(Note that $\lim_{m \to \infty} \beta_i^{\frac{1}{m}} = 1$ hence the interpolation points can be made as near to 0 as we want, so a solution surely exists for m large enough).

Step 2) If S(s) is an SBR function which then interpolates

$$S(\alpha_i) = \sigma_i = \frac{\beta_i^{\frac{1}{m}} - 1}{\beta_i^{\frac{1}{m}} + 1}$$

then from Fact 3.4

$$Z(s) = \frac{1 + S(s)}{1 - S(s)}$$

is a EP/SPR function which interpolates

$$Z(\alpha_i) = \beta_i^{\frac{1}{m}}$$

Step 3) From Fact 3.3 the function

$$U(s) = \left(Z(s) \right)^m$$

is a unit in H^∞ which interpolates to

$$U(\alpha_i) = \beta_i.$$

The above algorithm for interpolation with units was first reported in [10]. A refinement of this algorithm which interpolates directly the points β_i with a positive real function is presented in [16]. This algorithm has been found to yield lower order units than the original algorithm given in [41].

3.3 SOLUTION OF ROBUST STABILIZATION PROBLEMS VIA INTERPOLATION ALGORITHMS. EXAMPLES

Example 3.2

Find the largest value of r_0 such that there exists a compensator which stabilizes all plants in the class $C(p_0(s), r(s))$, where

$$p_0(s) = \frac{1}{(1-s)(2-s)} \; ; \; r(s) = r_0 > 0 \; .$$

<u>Solution :</u>

We have

$$B(s) = \frac{(1-s)(2-s)}{(1+s)(2+s)} \; ;$$

$$\tilde{p}_0(s) = B(s) \, p_0(s) = \frac{1}{(s+1)(s+2)} \; .$$

The equivalent interpolation problem is then :

Find an SBR function $u(s)$ which interpolates to

$$u(1) = \frac{r_m(1)}{\tilde{p}_0(1)} = 6 \, r_0; \qquad u(2) = \frac{r_m(2)}{\tilde{p}_0(2)} = 12 \, r_0 \, .$$

For a solution to exist the Pick matrix

$$
P = \begin{pmatrix} \dfrac{1 - 36\, r_0^2}{2} & \dfrac{1 - 72\, r_0^2}{3} \\[2mm] \dfrac{1 - 72\, r_0^2}{3} & \dfrac{1 - 144 r_0^2}{4} \end{pmatrix}
$$

should be positive definite. After some computations we find that the P matrix is positive definite if

$$r_0 < 0.04679 \ .$$

□

Example 3.3

Find a compensator which robustly stabilizes the nominal plant

$$p_0(s) = \frac{1}{s^2 - 2\,s + 2} \ ; $$

given the additive unstructured uncertainty bound $\quad r_m(s) = \dfrac{.25}{s + 1}\ .$

Solution :

We have

$$B(s) = \frac{s^2 - 2\,s + 2}{s^2 + 2\,s + 2} \ ; \qquad \tilde{p}_0(s) = B(s)\, p_0(s) = \frac{1}{s^2 + 2\,s + 2} \ ;$$

Interpolation conditions on the strictly bounded real function u(s) are

$$u(1 + j) = \frac{r_m(1+j)}{\tilde{p}_0(1+j)} = \frac{3 + j}{5} \ ;$$

and

$$u(1 - j) = \frac{r_m(1-j)}{\tilde{p}_0(1-j)} = \frac{3 - j}{5} \ .$$

Since $r_m(s)$ is of relative degree 1 we also need

$$u(\infty) = 0.$$

If we construct the Fenyves array for this interpolation problem, we have

1+j	1-j	∞	
$\frac{3+j}{5}$	$\frac{3-j}{5}$	0	
	0.338-j0.7077	-0.6-j0.2	
		-0.545+j0.731	

Note that all the elements in the array are in modulo less than one so from Theorem 3.2 the problem admits a solution.
We set

$$u_3(s) = -0.545 + j\, 0.731$$

thus we obtain the Schur function

$$u_1(s) = \frac{(2.2901-j1.0621)s + (-0.8588+j0.3983)}{(0.1789-j0.083)s^2 + (2.3616-j1.0952)s + (0.9303-j0.4315)} \; ;$$

To obtain an SBR function from the Schur function

$$u(s) = \frac{1}{2}\left(u_1(s) + \overline{u_1(\bar{s})} \right) = \frac{12.8\, s - 4.8}{s^2 + 13.2\, s + 5.2} \; ;$$

We then compute

$$q(s) = \frac{u(s)\, B(s)}{r_m(s)} = \frac{12.8\, s^4 - 17.6\, s^3 + 4.8\, s^2 + 25.6\, s - 9.6}{0.25\, s^4 + 3.8\, s^3 + 8.4\, s^2 + 9.2\, s + 2.6} \; ;$$

and finally the robustly stabilizing compensator

$$c(s) = \frac{q(s)}{1 - q(s)p_0(s)} = \frac{51.2\, (s + 1)(s - .375)}{(s + 16.2916)(s+0.9084)} \; .$$

□

Example 3.4

Consider the problem of example 2.10 of finding a stable compensator to stabilize the plant

$$p(s) = \frac{(s - 1)(s - 4)}{(s - 2)(s - 3)}.$$

To solve this problem we must find a unit U(s) which interpolates to

$$U(1) = \beta_1 = \frac{1}{2} ; \quad \text{and} \quad U(4) = \beta_2 = \frac{2}{25} ;$$

we convert this problem to that of finding an integer m such that the points

$$\sigma_i = \frac{\beta_1^{\frac{1}{m}} - 1}{\beta_1^{\frac{1}{m}} + 1} \quad , \quad i = 1, 2 ;$$

can be interpolated with an SBR function.
For m = 1 we have,

$$\sigma_1 = -0.33 ; \quad \sigma_2 = -0.8519 ;$$

and Fenyves array is given by

1	4	
-0.33	-0.8519	
	-1.2069	

thus for m = 1 Theorem 3.2 is violated and no solution exists.
For m = 2 we obtain

$$\sigma_1 = -0.1716 ; \quad \sigma_2 = -0.559 ;$$

and Fenyves array is given from

1	4	
-0.1716	-0.559	
	-0.7143	

In this case the problem admits solution (since all the diagonal terms have modulus less than 1).

From (1) with $u_2(s) = -0.7143$, we have

$$u_1(s) = \frac{-0.7891\ s + 0.4835}{s + 0.7817}\ ;$$

and

$$U(s) = \left(\frac{1 + u_1(s)}{1 - u_1(s)}\right)^2 = \frac{0.0139\ (s^2 + 6)}{(s + 0.1667)^2}\ ;$$

Thus the compensator is

$$c(s) = \frac{U(s) - D(s)}{N(s)} = \frac{-0.9861\ s^2 - 0.0694\ s + 0.0833}{s^2 + 0.333\ s + 0.0278}$$

□

Example 3.5

Consider the problem in example 2.11 of finding a compensator which simultaneously stabilizes the two plants

$$G_0(s) = \frac{1}{s + 1}\ ;\qquad G_1(s) = G_0(s) + \frac{(s - 1)(s - 3)}{(s - 2)^3}\ .$$

To solve this problem we need a unit U(s) which interpolates to

$$U(1) = \frac{1}{4}\ ;\qquad U(3) = \frac{1}{16}\ .$$

Analogously to the previous example we obtain for $m = 2$

$$u_1(s) = \frac{-0.8182\ s + 0.2727}{s + 0.6364}\ ;$$

and

$$U(s) = \frac{0.01\ (s + 5)^2}{(s + 0.2)^2}\ ;$$

We then obtain

$$R(s) = \frac{-0.99(s + 0.333)(s - 0.0909)}{(s + 0.2)^2} \; ;$$

and finally the compensator is given by

$$c(s) = \frac{R(s)}{1 - G_0(s)R(s)} = \frac{-0.99\,s^3 - 1.23\,s^2 - 0.21\,s + 0.03}{s^3 + 2.39\,s^2 + 0.68\,s + 0.01} \quad .$$

□

57

3.4 EXERCISES

1. Consider the problem of simultaneously stabilizing the two plants

$$G_0(s) = \frac{1}{s+1} \quad , \quad G_1(s) = \frac{s^3-s^2+172\ s-168}{(s+1)(s^2+173s-170)} \quad .$$

Use the theory in section 2.3 to reduce this to an interpolation problem with units in H^∞ (see problem 2 in section 2.5). Use the theory in section 3.2 to solve the resulting interpolation problem. Does there exist a second order unit which solves the interpolation problem ? Hint: A second order unit must be of the form

$$U(s) = \frac{a\ s^2+b\ s+c}{s^2+d\ s+e}$$

where a, b, c, d and e are all positive constants. The interpolation conditions then generate a system of linear equations in these constants. The issue then reduces to the question of a feasible solution to a linear programming problem.

2. We define a positive analytic (PA) function to be a function $Z(s)$ with the following properties
    (i) $Z(s)$ is analytic for Re $s \geq 0$;
    (ii) Re $Z(s) \geq 0$, for Re $s > 0$.
    Note that if $Z(s)$ is PA and real, then it is a <u>positive real</u> function, with no poles on the $j\omega$-axis.
    Show that if Re $\beta > 0$, then

$$S(s) = \frac{Z(s) - \beta}{Z(s) + \bar{\beta}}$$

is a Schur function if and only if $Z(s)$ is a PA function.

Hint: study $1-\bar{s}s$ and $z+\bar{z}$. Use this result to show that for the following mapping

$$\frac{z_i(s) - \xi_i}{z_i(s) + \bar{\xi}_i} = \left(\frac{s - \alpha_i}{s + \bar{\alpha}_i} \right) \left(\frac{z_{i+1}(s) - 1}{z_{i+1}(s) + 1} \right)$$

with $\mathrm{Re}\ \xi_i > 0$ and $\mathrm{Re}\ \alpha_i > 0$, we have:

(i) If $z_{i+1}(s)$ is PA, then so is $z_i(s)$. In addition $z_i(\alpha_i) = \xi_i$, for any $z_{i+1}(s)$.

(ii) If $z_i(s)$ is PA and $z_i(\alpha_i) = \xi_i$, then $z_{i+1}(s)$ is PA.

(iii) If $z(s)$ is a PA function which interpolates the point $z(\alpha_i) = \xi_i$, then

$$W(s) = \frac{1}{2} \left(z(s) + \overline{z(\bar{s})} \right)$$

is a PR function which interpolates the same points if $z(\bar{\alpha}_i) = \bar{z}(\alpha_i)$.

The above mapping may be used to directly interpolate with a positive real function and provides a simpler alternative to the mapping given in section 3.2. For more details an interpolation with positive real functions see Youla and Saito [44].

3.   Find a parameterization of all compensators c(s) which robustly stabilize the plant

$$p_0(s) = \frac{1 - s}{(2 - s)(1 + s)}$$

given the additive perturbation bound

$$|\delta p(j\omega)| < \frac{1}{20}, \text{ all } \omega.$$

Hint: use the non-uniqueness of the interpolation solution in section 3.1 to parameterize c(s) as follows

$$c(s) = \frac{A_1(s)\, U(s) + A_2(s)}{A_3(s)\, U(s) + A_4(s)};$$

where $U(s)$ is an <u>arbitrary</u> bounded real function and $A_1(s)$, $A_2(s)$, $A_3(s)$ and $A_4(s)$ may be taken as known polynomials. Find the polynomials $A_1(s)$, $A_2(s)$, $A_3(s)$ and $A_4(s)$ for the above problem.

CHAPTER 4

H^∞ SENSITIVITY/DISTURBANCE-REJECTION OPTIMIZATION

4.1 THE EQUIVALENT OPTIMAL INTERPOLATION PROBLEM

We use the notation and results in Zames and Francis [45] in this developments of the sensitivity/disturbance-rejection optimization problem. Thus as in [45], the compensator is now denoted F(s), and is placed in the feedback path rather than the forward path. These are only minor modifications from previous notation, but they expedite reference to the material in [45].

Consider the control feedback configuration of Fig. 4.1 .

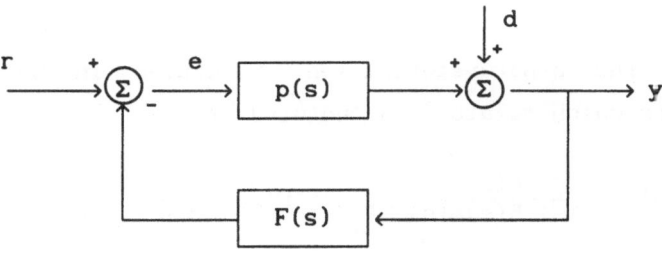

Fig. 4.1 Closed-loop system with compensator in the feedback loop.

Note that for this configuration the sensitivity function is identical to the transfer function between disturbance d and output y, i.e.

$$S = S_p^T = y/d = \frac{1}{1 + p(s)F(s)} \cdot$$

Thus if we attempt to design a compensator F(s) which minimizes some norm on the sensitivity function S(s), we will achieve both minimal sensitivity to plant perturbations and maximal disturbance rejection. In general we need to weight certain frequencies more than others, so that a weighting function $W_1(s)$ is introduced. Finally we select an H^∞-norm as a performance measure for two reasons. One is that the optimization then developed can be applied to the design of optimally robust systems, the other is that H^{∞} optimization does not require knowledge of the disturbance signal spectrum as does the H^2 optimization theory [42].

The optimization problem then is to find a compensator F(s) which minimizes

$$\|W_1(s) \ S(s)\|_{\infty} \tag{4.1}$$

while guaranteeing internal stability of the closed-loop system. Without loss of generality $W_1(s)$ is taken to be an H^{∞} function with no zeros in Re s $\geq$ 0. Note that here we assume that the plant is not perturbed and that p(s) represents the nominal plant transfer function. Thus designs which result from the above optimization, while minimizing a (nominal) sensitivity function, may not be robust with respect to finite plant perturbations.

Now recall the q-parameterization introduced in Sec. 2.1. The relationships which relate F(s) and q(s) are

$$q(s) = \frac{F(s)}{1 + F(s)p(s)} \quad , \quad F(s) = \frac{q(s)}{1 - q(s)p(s)} \ .$$

By this choice the problem is converted to one of finding an H^{∞} function q(s) such that satisfies internal stability conditions and minimizes

$$\|W_1(s) \ (\ 1 \ - \ p(s)q(s) \ )\|_{\infty} \ .$$

Let

$$u(s) = \frac{W_1(s) \ (\ 1 \ - \ p(s)q(s) \ )}{B_p(s)} \tag{4.2}$$

where $B_p(s)$ is the Blaschke product formed from the poles of p(s) in the RHP, i.e.

$$B_p(s) = \prod \left(\frac{\alpha_i - s}{\overline{\alpha}_i + s} \right)$$

where α_i are the RHP poles of p(s). Assume for simplicity that all the poles in the RHP of p(s) are distinct and that p(s) has no poles on the jω axes.

If we find an H^{∞} function u(s) with minimal norm which satisfies the interpolation conditions

$$u(\beta_1) = \frac{W_1(\beta_1)}{B_p(\beta_1)} \; ;$$

where β_1 are the RHP zeros of p(s), then the function, obtained from (4.2)

$$q(s) = \frac{1}{p(s)}\left(1 - \frac{B_p(s)\, u(s)}{W_1(s)} \right) \qquad (4.3)$$

solves our problem.

Proof

Blaschke products are inner functions, thus

$$\left\| W_1(s)\, (\, 1 - p(s)q(s)\,)\right\|_{\infty} = \left\| u(s)\right\|_{\infty} \; ;$$

so minimizing $\left\| W_1(s)\, S(s)\right\|_{\infty}$ is equivalent to minimizing $\left\| u(s)\right\|_{\infty}$.
Now remember the conditions on q for internal stability :

1) $q \in H^{\infty}$;
2) q must have zeros at the poles of p in the RHP;
3) pq must interpolate to 1 at the poles of p in the RHP .

From relation (4.3) it follows that conditions 2) and 3) are automatically satisfied.
The interpolation constraints on u(s) guarantee that the RHP zeros of p(s) in the above relation are cancelled from the other term so q(s) has no poles in the RHP.
Finally to satisfy 1) we must ensure that q(s) is a proper function.

In general the q(s) which minimizes (4.1) is not proper.

If the optimal $\tilde{q}(s)$ given by (2) is not a proper function, consider the sequence

$$\tilde{q}_n(s) = q_0(s) + \left(\tilde{q}(s) - q_0(s) \right) \left(\frac{n}{s+n} \right)^{k+l+1} \qquad (4.4)$$

where k is the relative degree of p(s), l is the relative degree of $W_1(s)$ and $q_0(s)$ is any stabilizing proper function q. It can be proven [45] that if $W_1(s)$ is strictly proper this sequence converges

$$\lim_{n \to \infty} \tilde{q}_n(s) = \tilde{q}(s) .$$

and

$$\lim_{n \to \infty} \| W_1(s) (1 - p(s) \tilde{q}_n(s)) \|_\infty = \| W_1(s) (1 - p(s) \tilde{q}(s)) \|_\infty ;$$

so, by a proper choice of n it is possible to find a proper $\tilde{q}_n(s)$ which yields a value of $\| W_1(s) S(s) \|_\infty$ which is as close to the optimal as we wish.

4.2 THE OPTIMAL INTERPOLATION SOLUTION

In this section we solve the problem of finding an H^∞ function u(s) of minimal H^∞ norm which interpolates to

$$u(\beta_i) = \vartheta_i; \qquad i=1,\ldots,\ell , \qquad \text{Re } \beta_i > 0.$$

Theorem 4.1

A BR function u(s) which interpolates $u(\beta_i) = \vartheta_i$; $i=1,\ldots,\ell$ has minimal norm equal to one if and only if :

$|\rho_k| = 1$ and $\rho_{k,j} = \rho_k$ for some $1 \le k \le \ell$, where $\rho_{k,j}$ are all terms of the k row in the Fenyves array.

A proof of theorem 4.1 may be found in Walsh [40], section 10.3.
Recall from the Nevanlinna-Pick theory that, from $|\rho_{kk}| = 1$ it follows that $u_k(s) \equiv \rho_k$ and that as a consequence if a solution u(s) exists, it must be an inner function

$$|u(j\omega)| \equiv 1 \qquad \forall \omega \in \mathbb{R} \ .$$

Suppose now that u(s) is an H^∞ function which interpolates to

$$u(\beta_i) = \vartheta_i; \quad i=1,\ldots,\ell$$

and has a minimal H^∞ norm equal to M

$$\|u(s)\|_\infty = M \ . \qquad (4.5)$$

Then the function $u_M(s) = \dfrac{u(s)}{M}$ which interpolates to

$$u_M(\beta_i) = \frac{\vartheta_i}{M} \quad i=1,\ldots,\ell$$

has a minimal norm equal to one. From theorem 4.1, in the Fenyves array of this function there exists a row k where

$$|\rho_k| = 1 \qquad \text{and} \qquad \rho_{k,j} = \rho_k \ .$$

From this consideration we can state the following algorithm to find a solution to the optimal interpolation problem :

Step 1) Compute row by row the elements of the Fenyves array, starting with the first row given by

$$\rho_1(M) = \frac{\beta_1}{M} \quad ; \quad \rho_{1,j}(M) = \frac{\beta_j}{M} \quad j = 2,\ldots.\ell \ .$$

α_1	α_2	α_3	.....	α_ℓ	
$\rho_1(M)$	$\rho_{1,2}(M)$	$\rho_{1,3}(M)$		$\rho_{1,\ell}(M)$	$v_1(s)$
	$\rho_2(M)$	$\rho_{2,3}(M)$		$\rho_{2,\ell}(M)$	$v_2(s)$
		....			.....
				$\rho_\ell(M)$	$v_\ell(s)$

where an initial arbitrary choice is made for M.

Step 2) If
$|\rho_k|=1$, for some row k, go to step 3. If $|\rho_i|<1$ for all i decrease M and go to step 1. If $|\rho_k|>1$ for some k increase M and go back to step 1.

Step 3) Let $v_k(s) \equiv \rho_k$, and compute the other functions $v_{k-1}(s), v_{k-2}(s), .., v_1(s)$.

Step 4) A solution to the optimal interpolation problem is then given by

$$u(s) = M v_1(s) \qquad (4.6)$$

with an optimal H^∞ norm equal to M

$$\|u(s)\|_\infty = M .$$

Note that the order of the interpolating function depends on the row k in step 2) for which we find a solution.
For problems with a small number of interpolation points one can study the condition $|\rho_k(M)|=1$ analytically to compute the value of M, as in the next example.

Example 4.1

Find a compensator $F(s)$ which minimizes $\|W_1(s)S(s)\|_\infty$ given

$$p(s) = \frac{(1 - s)(4 - s)}{(2 - s)(3 - s)} \; ; \qquad W_1(s) = \frac{s + 1}{8s + 1} \; .$$

Solution :

$$B_p(s) = \frac{(2 - s)(3 - s)}{(2 + s)(3 + s)} \; ;$$

The interpolation points are

$$u(1) = \frac{W_1(1)}{B_p(1)} = \frac{4}{3} \; ; \qquad u(4) = \frac{W_1(4)}{B_p(4)} = \frac{35}{11} \; ;$$

The Fenyves array is

$$
\begin{array}{cc|}
1 & 4 \\
\hline
\dfrac{4}{3\,M} & \dfrac{35}{11\,M} \\
& \\
& \rho_2(M)
\end{array}
$$

where

$$\rho_2(M) = \frac{5}{3}\;\frac{\dfrac{35}{11\,M} - \dfrac{4}{3\,M}}{1 - \dfrac{4}{3\,M}\dfrac{35}{11\,M}} \; ;$$

From $\quad \rho_2(M) = 1 \quad$ we obtain as optimal norm the value

$$M = 4.1124.$$

thus

$$v_1(s) = \frac{1.3242\,s - 0.6758}{1.3242\,s + 0.6758} \; ;$$

and

$$u(s) = M\,v_1(s) = 4.1124\left(\frac{1.3242\,s - 0.6758}{1.3242\,s + 0.6758}\right) ;$$

Substituting u(s) in (2) we obtain q(s) and then the compensator F(s) as

$$F(s) = \frac{q(s)}{1 - p(s)q(s)} = \frac{2.4317(-3.0323\ s^2 + 16.9208\ s + 81.0991\ )}{(8\ s + 1)\ (1.3242\ s - 0.6758)}$$

□

Note that the optimal compensator in example 4.1 is unstable. This is not a special case. Indeed since the optimal u(s) is all pass, with zeros all in the RHP, and the compensator is given by (use 4.1 and F = q/(1-qp))

$$F(s) = \frac{W_1(s) - B(s)u(s)}{p(s)\ B_p(s)\ u(s)}\ ,$$

it is obvious that unless u(s) is a constant, i.e. p(s) has only one zero in the RHP, the RHP zeros of u(s) will be unstable poles of F(s).

Example 4.2
Find a sequence of controller $F_n(s)$ such that the minimum of $\|W(s)S(s)\|_\infty$ is approached, given

$$p(s) = \frac{1 - s}{(1 + s)^2}\ , \qquad W_1(s) = \frac{1}{s + 1}\ ;$$

Solution :
Since p(s) is stable $B_p(s) \equiv 1$ and the only interpolation condition is

$$u(1) = \frac{W_1(s)}{B_p(1)} = \frac{1}{2}\ ;$$

we pick

$$u(s) = \frac{1}{2}\ .$$

NOTE : For a single interpolation point, a constant is optimal. Then the optimal q(s) is

$$\tilde{q}(s) = - \frac{1 - \frac{u(s)}{W_1(s)}}{p(s)} = \frac{1}{2} (s + 1)^2 .$$

Since $W(s)$ and $p(s)$ are of relative degree 1 each, we need the sequence of proper $q(s)$ given by

$$\tilde{q}_n(s) = q_0(s) + \left(\tilde{q}(s) - q_0(s) \right) \left(\frac{n}{s + n} \right)^3 ;$$

where $q_0(s)$ is any strictly proper H^∞ function which stabilizes the nominal closed-loop system. Since in this case $p(s)$ is itself stable $q_0(s)$ need not satisfy any interpolation conditions.
A simple choice is

$$q_0(s) = \frac{1}{s + 1} ;$$

then

$$\tilde{q}_n(s) = \frac{1}{s + 1} + \left(\frac{1}{2}(s + 1)^2 - \frac{1}{s + 1} \right) \left(\frac{n}{s + n} \right)^3 ;$$

and

$$\tilde{F}_n(s) = \frac{\tilde{q}_n(s)}{1 - p(s)\, \tilde{q}_n(s)} ;$$

is the required compensator sequence.

□

4.3 EXERCISES

1. Consider the nominal plant

$$p(s) = \frac{1 - s}{(2 - s)(1 + s)} \ .$$

Compute

$$\inf \left\| W_1(s) \ S(s) \right\|_\infty$$

given

$$W_1(s) = \frac{1}{s + 1} \ .$$

Find the (improper) compensator F(s) which yields the above inf. What are the implications of F(s) being improper ? Find a proper compensator corresponding to n=10 in equation (4.3). What is the value of $\left\| W_1(s) \ S(s) \right\|_\infty$, if the proper compensator just computed is used in the feedback loop ?

2. By q-parameterization, scalar optimal H^∞ control problems can always be reduced to the following optimization problem

$$\inf_{Q(s) \in H^\infty} \left\| T_1(s) - T_2(s) \ Q(s) \right\|_\infty$$

where $T_1(s)$ and $T_2(s)$ are H^∞ functions.
Show that this can be reduced to the optimal interpolation problem discussed in section 4.2. Hint: Let $U = T_1 - T_2 Q$, and compute the values that U must interpolate if Q is to belong to H^∞. Find Q(s) and the minimal H^∞ norm via optimal interpolation for the data

$$T_1(s) = \frac{s + 1}{10s + 1} \ , \quad T_2(s) = \frac{(s - 1)(s - 2)}{(s + 1)^2} \ .$$

3. Since the solution to the optimal interpolation problem is always an all-pass function it follows that the optimal sensitivity frequency response must satisfy

$$|S(j\omega)| = \frac{M}{|W_1(j\omega)|} \;, \text{ all } \omega.$$

Given the data

$$W_1(s) = \frac{s + 1}{10s + 1} \;, \; p(s) = \frac{s^2 - s + 1}{s^2 + s + 1}$$

compute M, and sketch $|S(j\omega)|$ versus ω.

Chapter 5
THE MATRIX H$^\infty$ SENSITIVITY OPTIMIZATION PROBLEM

5.1 MATRIX Q - PARAMETERIZATION

In this chapter we extend the sensitivity optimization problem developed in chapter 4 for single-input-single-output systems to multivariable systems. It is convenient in the matrix case to introduce a Q-parameterization, as in chapter 2, which does not require interpolation methods. Following [43] and [12] all matrix compensators which yield a nominally internally stable feedback system are parameterized in terms of an arbitrary matrix Q(s) whose entries are all H$^\infty$ functions. Following the notation in [37], the set of matrices with all elements in H$^\infty$ is denoted M(H$^\infty$). It is then shown that the matrix H$^\infty$ sensitivity problem can then be reduced the so-called Nehari approximation problem. In chapter 6 a solution to the Nehari problem via Hankel norms is outlined, together with some recent 2-Riccati-equation approach to the H$^\infty$ control problem. The theory developed for H$^\infty$ sensitivity optimization in chapter 5 and 6, can also be applied to robust stabilization problems as outlined in chapter 7.

Definition 5.1 [37]

Given a matrix transfer function p(s) of a plant an ordered pair $(n_r(s), d_r(s))$, where $n_r(s), d_r(s) \in$ M(H$^\infty$), represents a right-coprime factorization of p(s) if:

1) $d_r(s)$ is square and $|d_r| \neq 0$, where $|d_r|$ denotes the determinant of d_r;
2) $p(s) = n_r(s)\, d_r^{-1}(s)$;
3) $n_r(s)$ and $d_r(s)$ are right coprime .

◻

Fact 5.1 [37]

$n_r(s)$ and $d_r(s)$ are right coprime if and only if there exists H^∞ matrices $u_r(s)$ and $v_r(s)$ that satisfy the following equation

$$u_r(s)\ n_r(s)\ +\ v_r(s)\ d_r(s)\ =\ I$$

(Right Matrix Bezout identity).

Similarly for left coprime factorizations the existence of H^∞ matrices $u_\ell(s)$ and $v_\ell(s)$ which satisfy the identity (left Bezout)

$$n_\ell(s)\ u_\ell(s)\ +\ d_\ell(s)\ v_\ell(s)\ =\ I$$

is necessary and sufficient for $n_\ell(s)$, $d_\ell(s)$ to be left coprime, with

$$p(s)\ =\ d_\ell^{-1}(s)\ n_\ell(s).$$

For computation of a doubly coprime fractional representation given a minimal realization of $p(s) = C(sI-A)^{-1}B + D$ we have the following theorem [29].

Theorem 5.1

Suppose $p(s) = C(sI-A)^{-1}B + D$ where $A \in \mathbb{R}^{n \times n}$, $B \in \mathbb{R}^{n \times m}$, $C \in \mathbb{R}^{p \times n}$, $D \in \mathbb{R}^{p \times m}$, (C,A) is detectable and (A,B) is stabilizable.

Select $K \in \mathbb{R}^{m \times n}$, $F \in \mathbb{R}^{n \times p}$ such that $A-BK$ and $A-FC$ are stable matrices (all eigenvalues have negative real parts). Define :

$$n_r(s) = (C-EK)(sI-A+BK)^{-1}B + E \quad ;\quad d_r(s) = I - K(sI-A+BK)^{-1}B \qquad ;$$

$$u_r(s) = K(sI-A+FC)^{-1}F \qquad\qquad ;\quad v_r(s) = I + K(sI-A+FC)^{-1}(B-FE)\ ;$$

$$d_\ell(s) = I - C(sI-A+FC)^{-1}F \qquad\quad ;\quad n_\ell(s) = C(sI-A+FC)^{-1}(B-FE) + E\ ;$$

$$v_\ell(s) = I + (C-EK)(sI-A+BK)^{-1}F \quad ;\quad u_\ell(s) = K(sI-A+BK)^{-1}F \qquad\qquad ;$$

Then

   i) all the eight matrices above defined belongs to $M(H^\infty)$;

  ii) $d_r(s)$ and $d_\ell(s)$ are nonsingular;

iii) $p(s) = n_r(s) \, d_r^{-1}(s) = d_\ell^{-1}(s) \, n_\ell(s)$;

iv)
$$\begin{bmatrix} v_r(s) & u_r(s) \\ -n_\ell(s) & d_\ell(s) \end{bmatrix} \begin{bmatrix} d_r(s) & -u_\ell(s) \\ n_r(s) & v_\ell(s) \end{bmatrix} = \begin{bmatrix} I & 0 \\ 0 & I \end{bmatrix} . \qquad (5.1)$$

Then it is possible to parameterize the set of all compensators $c(s)$ that stabilize $p(s)$ in terms of an arbitrary H^∞ matrix $Q(s)$ as follows [37]

$$c(s) = (Q(s) \, n_\ell(s) + v_r(s))^{-1}(-Q(s) \, d_\ell(s) + u_r(s)) ; \qquad (5.2)$$

where

$$Q(s) \in M(H^\infty), \; |Q(s) \, n_\ell(s) + v_r(s)| \neq 0 .$$

This choice guarantees that all terms in h matrix, see equation (2.1), are affine (linear plus a constant) in the Q parameter and that the closed loop plant is stable.

Consider for example the sensitivity matrix

$$S = (I + c \, p)^{-1}$$

by substituting $c(s)$ in this expression, after some algebraic manipulations, we have

$$I + cp = I + (Q \, n_\ell + v_r)^{-1} d_r^{-1}$$

So, we can write

$$S = (I + cp)^{-1} = d_r (Q \, n_\ell - v_r)$$

In this expression we can see that S is affine in the Q parameter.

The general expression which represent matrices affine in Q is

$$T_1 + T_2 Q T_3 \qquad (5.3)$$

where $T_1, T_2, T_3, Q \in M(H^\infty)$.
For example for the sensitivity matrix S, we have

$$T_1 = -d_r v_r; \quad T_2 = d_r ; \quad T_3 = n_\ell .$$

5.2 MATRIX INNER/OUTER FACTORIZATION

Similarly to the scalar case a stable transfer matrix can be written as a product of an inner matrix times an outer matrix.

Definition 5.2
A square transfer matrix $M(s) \in M(H^\infty)$ is *outer* if $M^{-1}(s)$ has no finite unstable poles.

Definition 5.3
A transfer matrix $\Theta(s) \in M(H^\infty)$ of dimension m x n is *inner* if satisfies

$$\Theta^T(-s)\Theta(s) = I \; \forall \, s \; , \; \text{if } m \geq n \text{ (row type)}$$

or

$$\Theta(s)\Theta^T(-s) = I \; \forall \, s \; , \; \text{if } m < n \text{ (column type)}.$$

Moreover when $m \neq n$, $\Theta(s)$ has a *complementary inner* $\Theta^\perp(s)$ such that

$$\begin{bmatrix} \Theta(s) & \Theta^\perp(s) \end{bmatrix} \quad \text{or} \quad \begin{bmatrix} \Theta(s) \\ \Theta^\perp(s) \end{bmatrix} \quad \text{is square and inner.}$$

An *inner-outer factorization* of a matrix $G(s) \in M(H^{\infty})$ of dimension $m \times n$ where $m \geq n$ (tall matrix) , is a factorization such that

$$G(s) = \left[\begin{array}{cc} \theta(s) & \theta^{\perp}(s) \end{array} \right] \left[\begin{array}{c} M(s) \\ 0 \end{array} \right] = \theta(s) \, M(s) \qquad (5.4)$$

where $\theta(s)$ is an inner matrix and $M(s)$ is an outer matrix.

For the case of $m < n$ (fat matrix) the same result can be applied to $G^{T}(s)$ by transposing the above formula

$$G(s) = \left[\begin{array}{cc} \tilde{M}(s) & 0 \end{array} \right] \left[\begin{array}{c} \tilde{\theta}(s) \\ \tilde{\theta}^{\perp}(s) \end{array} \right] = \tilde{M}(s) \, \tilde{\theta}(s) \qquad (5.5)$$

State-space formulas for the computation of matrix inner/outer factors are given in Doyle [19]. These computations are reduced to the solution of Riccati equations.

5.3 REDUCTION OF THE OPTIMAL H^∞ CONTROL PROBLEM TO A MATRIX NEHARI PROBLEM

Many H^∞ control problems can be reduced to the following H^∞ optimization problem [19]:

Find a matrix $Q(s) \in M(H^\infty)$ which minimizes

$$\| T_1(s) + T_2(s) \, Q(s) \, T_3(s) \|_\infty \tag{5.6}$$

where $T_1(s)$, $T_2(s)$ and $T_3(s) \in M(H^\infty)$ are given rational matrices. In this section it will be shown how this problem can be reduced to an equivalent Nehari problem. The Nehari problem, which will be defined more precisely shortly, is basically a problem of approximating a stable matrix by a totally unstable matrix.

If T_2 and T_3 are both square then may be factored as

$$T_2(s) = \theta_2(s) \, M_2(s) \; ;$$
$$T_3(s) = M_3(s) \, \theta_3(s) \; ;$$

where θ_1 and θ_2 are square and inner and M_1 and M_2 are square and outer. Thus

$$T_1(s) + T_2(s) \, Q(s) \, T_3(s) = T_1(s) + \theta_2(s) \, M_2(s) \, Q(s) \, M_3(s) \, \theta_3(s) =$$
$$= T_1(s) + \theta_2(s) \, \tilde{Q}(s) \, \theta_3(s) \; ;$$

where

$$\tilde{Q}(s) = M_2(s) \, Q(s) \, M_3(s) \; .$$

Now by taking the H^∞ norm

$$\| T_1 + T_2 \, Q \, T_3 \|_\infty = \| T_1 + \theta_2 \, \tilde{Q} \, \theta_3 \|_\infty = \| \theta_2 (\, \theta_2^* \, T_1 \, \theta_3^* + \tilde{Q} \,) \, \theta_3 \|_\infty =$$
$$= \| T_{11} + \tilde{Q} \|_\infty \; .$$

where $T_{11} = \theta_2^* \, T_1 \, \theta_3^*$.

The optimal H^∞ control problem is then reduced to an equivalent so-called 1 Block problem

$$\inf_{\tilde{Q} \in M(H^\infty)} \| \, T_{11} + \tilde{Q} \, \|_\infty$$

□

If T_2 or T_3 are not square more computation is needed. For example when T_2 is tall and T_3 fat

$$T_1 + T_2 \, Q \, T_3 = T_1 + \boxed{T_2} \; Q \; \boxed{ T_3 }$$

as was shown is Sec.5.2, T_2 and T_3 can be factored as

$$T_2(s) = \begin{bmatrix} \theta_2 & \theta_2^\perp \end{bmatrix} \begin{bmatrix} M_2 \\ 0 \end{bmatrix} ; \quad T_3(s) = \begin{bmatrix} M_3 & 0 \end{bmatrix} \begin{bmatrix} \tilde{\theta}_3 \\ \tilde{\theta}_3^\perp \end{bmatrix} .$$

Thus,

$$T_1 + T_2 \, Q \, T_3 = T_1 + \begin{bmatrix} \theta_2 & \theta_2^\perp \end{bmatrix} \begin{bmatrix} M_2 \\ 0 \end{bmatrix} Q \begin{bmatrix} M_3 & 0 \end{bmatrix} \begin{bmatrix} \tilde{\theta}_3 \\ \tilde{\theta}_3^\perp \end{bmatrix} =$$

$$= T_1 + \begin{bmatrix} \theta_2 & \theta_2^\perp \end{bmatrix} \begin{bmatrix} M_2 Q \, M_3 & 0 \\ 0 & 0 \end{bmatrix} \begin{bmatrix} \tilde{\theta}_3 \\ \tilde{\theta}_3^\perp \end{bmatrix} =$$

$$= \begin{bmatrix} \theta_2 & \theta_2^\perp \end{bmatrix} \left(\begin{bmatrix} \theta_2 & \theta_2^\perp \end{bmatrix}^* T_1 \begin{bmatrix} \tilde{\theta}_3 \\ \tilde{\theta}_3^\perp \end{bmatrix}^* + \begin{bmatrix} \tilde{Q} & 0 \\ 0 & 0 \end{bmatrix} \right) \begin{bmatrix} \tilde{\theta}_3 \\ \tilde{\theta}_3^\perp \end{bmatrix}$$

where

$$\tilde{Q} = M_2 \, Q \, M_3 ; \quad \begin{bmatrix} \theta_2(s) & \theta_2^\perp(s) \end{bmatrix}^* = \begin{bmatrix} \theta_2(-s) & \theta_2^\perp(-s) \end{bmatrix}^T ;$$

Now by taking the H^{∞} norm

$$\| T_1 + T_2 Q T_3 \|_{\infty} = \left\| \begin{bmatrix} T_{11} & T_{12} \\ T_{21} & T_{22} \end{bmatrix} + \begin{bmatrix} \tilde{Q} & 0 \\ 0 & 0 \end{bmatrix} \right\|_{\infty}$$

where

$$\begin{bmatrix} T_{11} & T_{12} \\ T_{21} & T_{22} \end{bmatrix} = \begin{bmatrix} \theta_2 & \theta_2^{\perp} \end{bmatrix}^* T_1 \begin{bmatrix} \tilde{\theta}_3 \\ \tilde{\theta}_3^{\perp} \end{bmatrix}^* ;$$

Our optimization problem is then reduced to the so-called 4 Block problem

$$\inf_{Q \in M(H^{\infty})} \| T_1 + T_2 Q T_3 \|_{\infty} = \inf_{\tilde{Q} \in M(H^{\infty})} \left\| \begin{matrix} T_{11} + \tilde{Q} & T_{12} \\ T_{21} & T_{22} \end{matrix} \right\|_{\infty} ; \qquad (5.7)$$

where

$$Q = M_2^{-1} \tilde{Q} M_3^{-1} .$$

It can be shown [22] that the 4 Block problem can be reduced to an equivalent 1 Block problem

$$\inf_{\tilde{Q} \in M(H^{\infty})} \| T + \tilde{Q} \|_{\infty} . \qquad (5.8)$$

Definition 5.4

We define the L^{∞} norm of a matrix transfer function H(s) analytic on the $j\omega$ axes as

$$\| H(s) \|_{L^{\infty}} = \sup_{\omega} \bar{\sigma} \left[H(j\omega) \right]$$

Note that the L^{∞} norm can be defined also for unstable plants. The only requirement is that H(s) should be analytic on the $j\omega$ axes.

Example 5.1

Consider the transfer function

$$G(s) = \frac{1}{s - 1} \; ;$$

G(s) is not an H^∞ function so we can not compute its H^∞ norm.
However it is possible to compute the L^∞ norm of G(s), i.e.

$$\|G(s)\|_{L^\infty} = \sup_\omega \frac{1}{\sqrt{\omega^2 + 1}} = 1.$$

□

Fact 5.2

Given a matrix $H(s) \in M(H^\infty)$ we have

$$\|H(s)\|_\infty = \|H(s)^*\|_{L^\infty}$$

where

$$H(s)^* = H^T(-s)$$

Note that if $H(s) \in M(H^\infty)$ then $H(s)^*$ is a *totally unstable* or *anticausal* matrix.

By using Fact 5.2 we reduce the 1 Block problem to the so-called Nehari approximation problem

$$\inf_{\tilde{Q} \in M(H^\infty)} \| T + \tilde{Q} \|_\infty = \inf_{\tilde{Q} \in M(H^\infty)} \| T^* + \tilde{Q}^* \|_{L^\infty} = \inf_{Y \in M(H^\infty)} \| G - Y^* \|_{L^\infty} \quad (5.9)$$

where

$$G(s) = T^*(s) \quad \text{and} \quad Y(s) = -\tilde{Q}(s).$$

Note that T(s) was totally unstable so $G(s) \in M(H^\infty)$. Thus the H^∞ optimization problem is reduced to a problem of the best approximation of a stable matrix, G(s), by a totally unstable matrix $Y^*(s)$ using an L^∞ matrix norm. See [33] for more details on this reduction.

We discuss the solution of this problem in the next chapter.

SOLUTION OF THE MATRIX H^∞ CONTROL PROBLEM

In this chapter we outline two approaches to the solution of the H^∞ control problem. In the first approach, due to Glover [23], Hankel-norm methods are used to solve the Nehari problem. This approach involves the solution of matrix Lyapunov equations, and Q-parameterization, which in turn requires the solution of other Lyapunov and Riccati equations. The total number of matrix equations to solved is large. In contrast the second approach, due to Doyle, Glover, Khargonekar, and Francis [20], involves the solution of only two Riccati equations, as in the optimal LQG problem [21].

6.1 THE HANKEL NORM APPROACH

Definition 6.1
Consider a minimal state space realization of a transfer matrix $G(s) \in M(H^\infty)$ with an asymptotically stable A matrix

$$G(s) = C(sI - A)^{-1}B + D := \begin{bmatrix} A & B \\ C & D \end{bmatrix}. \qquad (6.1)$$

The assignment symbol := is used to conveniently exhibit the realization matrices A, B, C and D.

We define the *Hankel singular values* of G(s) as

$$\sigma_i(G(s)) \equiv \left[\lambda_i(P \, Q \,) \right]^{1/2}; \qquad (6.2)$$

where P and Q are the Controllability and the Observability Gramians respectively, defined as follows

$$P = \int_0^\infty e^{At} BB^T e^{A^T t} dt \qquad (6.3)$$

$$Q = \int_0^\infty e^{A^T t} C^T C e^{At} dt \qquad (6.4)$$

It is easily verified that these Gramians satisfy the following Lyapunov equations:

$$A P + P A^T + B B^T = 0 \qquad (6.5)$$
$$A^T Q + Q A + C^T C = 0 \qquad (6.6)$$

Note : A realization is said to be an <u>ordered balanced</u> when the solution P and Q of the above Lyapunov equations are equal and diagonal, i.e.

$$P = Q = \Sigma = \mathrm{diag}(\sigma_1, \sigma_2, \ldots\ldots, \sigma_n) \qquad (6.7)$$

with

$$\bar{\sigma} = \sigma_1 \geq \sigma_2 \geq \ldots\ldots \geq \sigma_n = \underline{\sigma}$$

$\square$

Definition 6.2

Let $G(s) = C (sI - A)^{-1}B + D \in M(H^\infty)$; then the *Hankel norm* of $G(s)$ is defined as

$$\|G(s)\|_H = \bar{\sigma}\left(G(s) \right) \qquad (6.8)$$

(the greatest Hankel singular value).

$\square$

The following theorem relates the Nehari problem to Hankel norms.

Theorem 6.1 [23]

Let $G(s) \in M(H^\infty)$; then

$$\inf_{Y \in M(H^\infty)} \| G - Y^* \|_{L^\infty} = \|G(s)\|_H . \qquad (6.9)$$

An explicit parameterization of all solutions $Y(s)$ satisfying the above relation can be found in [23], $Y(s)$ then relates back to the computation of $Q(s)$ and hence $c(s)$.

6.2 H^∞ CONTROL WITH STATE FEEDBACK: 1-RICCATI EQUATION SOLUTION

Consider the linear time-invariant system described by the state space equations

$$\begin{cases} \dot{x} = A\,x + B_1 w + B_2\,u \\ z = C_1 x + \qquad\quad D_{12} u \end{cases} \tag{6.10}$$

where u is the control input, w is the disturbance vector and z is the controlled output.

In this section we consider the problem of finding a constant matrix F such that the state feedback control law

$$u = F\,x$$

stabilizes the system and that

$$\|T_{zw}(s)\|_\infty \le 1 \tag{6.11}$$

where $T_{zw}(s)$ is the closed-loop transfer matrix between w and z.

Before explaining the solution to this problem some related results will be summarized.

Lemma 6.1

Let $G(s) = C(sI - A)^{-1}B$; if there exists a real positive definite symmetric matrix X solution to the Riccati equation

$$A^T X + X\,A + C^T C + X\,B\,B^T\,X = 0 \tag{6.12}$$

and the pair (A, C) is observable, then

$$\|G(s)\|_\infty \le 1. \tag{6.13}$$

Proof

From Lyapunov stability theory if X is positive definite and (A,C) is observable, then A must be an asymptotically stable matrix.

We show next that the matrix

$$I - G^T(j\omega)G(j\omega)$$

is non-negative definite which then completes the proof of (6.13), since then we have $\|G(j\omega)\|_\infty \le 1$ all ω.

Let us consider equation (6.12) written as

$$-A^TX - X A = C^TC + X B B^TX$$

then add and subtract sX

$$(-sI - A^T)X + X(sI - A) = C^TC + X B B^TX.$$

Let

$$\Phi(s) = (sI - A)^{-1}$$

and pre-multiply by $B^T\Phi^*$ and post-multiply by ΦB

$$B^TX \Phi B + B^T\Phi^* X B = B^T\Phi^*C^TC \Phi B + B^T\Phi^*X B B^TX \Phi B$$

add I to both sides and rearrange as

$$I - G^*G = I - B^TX \Phi B - B^T\Phi X B + B^T\Phi^*X B B^T X \Phi B$$

(remember that $G = C \Phi B$)

then for $s=j\omega$ we have

$$I - G^*G = (I - B^T\Phi X B)^*(I - B^T \Phi X B) \ge 0$$

□

Lemma 6.2

Let $G(s) = C(sI - A)^{-1}B$; if

$$\|G(s)\|_\infty < 1$$

then the Hamiltonian matrix H associated with the Riccati equation

$$H = \left(\begin{array}{c|c} A & B\,B^T \\ \hline -C^T C & -A^T \end{array}\right) \qquad (6.14)$$

cannot have any $j\omega$-axis eigenvalues.

Proof

Consider the realization

$$G(s) := \left[\begin{array}{cc} A & B \\ C & 0 \end{array}\right]$$

then

$$G^*(s) = G^T(-s) = B^T(-sI - A^T)^{-1}C^T = -B^T(sI - (-A^T))^{-1}C^T$$

$$G^*(s) := \left[\begin{array}{cc} -A^T & C \\ -B^T & 0 \end{array}\right]$$

Now to compute a state space description of the following product

$$G^*(s)G(s)$$

consider this as the cascade of the two systems $G(s)$ and $G^*(s)$

combining these equations, we get

$$\begin{cases} \dot{x} = A\,x + B\,u \\ \dot{\xi} = C^T C\,x - A^T \xi \\ w = -\,B^T \xi \end{cases}$$

which is equivalent to

$$G^*(s)G(s) := \left[\begin{array}{cc|c} A & 0 & B \\ C^T C & -A^T & 0 \\ \hline 0 & -B^T & 0 \end{array}\right] \quad .$$

Then

$$I - G^*(s)G(s) := \left[\begin{array}{cc|c} A & 0 & B \\ C^T C & -A^T & 0 \\ \hline 0 & B^T & I \end{array}\right] \qquad (6.15)$$

Now to evaluate $\left(I - G^*(s)G(s) \right)^{-1}$ we need another result.

Given a linear system

$$M(s) := \left[\begin{array}{cc} A_M & B_M \\ C_M & D_M \end{array}\right]$$

assume that D_M^{-1} exists; then

$$\begin{cases} \dot{x} = A_M\,x + B_M\,u \\ y = C_M\,x + D_M\,u \end{cases}$$

the inverse system (y input - u output) is given by

$$\begin{cases} \dot{x} = (A_M - B_M D_M^{-1} C_M)\,x + B_M D_M^{-1}\,y \\ u = D_M^{-1} y - D_M^{-1} C_M\,x \end{cases}$$

so we have

$$M^{-1}(s) := \begin{bmatrix} A_M - B_M D_M^{-1} C_M & B_M \\ -D_M^{-1} C_M & D_M^{-1} \end{bmatrix}$$

Then applying this last formula to equation (6.15) we obtain the realization

$$W(s) = \left(I - G^*(s)G(s) \right)^{-1} := \left[\begin{array}{cc|c} A & BB^T & B \\ C^T C & -A^T & 0 \\ \hline 0 & B^T & I \end{array} \right] \qquad (6.16)$$

If $\|G(s)\|_\infty < 1$ then $I - G^*(j\omega)G(j\omega) > 0$, $\forall \, \omega$, and hence

$$W(s) \in M(L^\infty).$$

So $W(s)$ has no poles on the imaginary axes. From (6.16) it is seen that the poles of $W(s)$ are the eigenvalues of H, hence H cannot have any $j\omega$-axis eigenvalues.

□

Consider the closed loop plant T_{zw}

$$\begin{cases} \dot{x} = (A + B_2 F) \, x + B_1 w \\ z = (C_1 + D_{12} F) \, x \end{cases}$$

from Lemma 6.1 $\|T_{zw}\|_\infty \leq 1$, if there exists a positive definite matrix X which satisfies

$$(A + B_2 F)^T X + X (A + B_2 F) + (C_1 + D_{12} F)^T (C_1 + D_{12} F) + X B_1 B_1^T X = 0.$$

Now let

$$\Sigma = D_{12}^T D_{12} ; \qquad \Theta = D_{12}^T C_1 + B_2^T X ;$$

then

$$A^T X + X A + C_1^T C_1 - \Theta^T \Sigma^{-1} \Theta + X B_1 B_1^T X + (F + \Sigma^{-1}\Theta)^T \Sigma (F + \Sigma \Theta) = 0.$$

Now assume

$$F = - \Sigma^{-1} \Theta$$

then we have

$$A^T X + X A + C_1^T C_1 - \Theta \Sigma^{-1} \Theta + X B_1 B_1^T X = 0 .$$

To simplify the results the following additional assumptions are made on $G(s)$

$$D_{12}^T D_{12} = I ; \qquad\qquad (6.17)$$

and

$$D_{12}^T C_1 = 0 . \qquad\qquad (6.18)$$

These assumptions are not essentially, but they considerable simplify the resulting equations.

Under the above conditions the equation becomes

$$A^T X + X A + C_1^T C_1 + X (B_1 B_1^T - B_2 B_2^T) X = 0. \qquad (6.19)$$

To find a state feedback compensator such that $\|T_{zw}\|_\infty \leq 1$ it is then necessary to find a positive definite solution of (6.19); then the "state-feedback" compensator is given by

$$u = F x$$

where

$$F = - B_2^T X$$

where X is a solution to the Riccati equation (6.19). Note that the Riccati equation (6.19) is not the standard LQ Riccati equation (section 1.1) since $B_1 B_1^T - B_2 B_2^T$ may be indefinite.

Note that :

- The Hamiltonian matrix associated with the Riccati equation (6.19) is

$$H = \left(\begin{array}{c|c} A & B_1 B_1^T - B_2 B_2^T \\ \hline -C_1^T C_1 & -A^T \end{array} \right) . \qquad (6.20)$$

If H has $j\omega$-axis eigenvalues, then from Lemma 6.2 our problem does not admit a solution.

- The special assumptions (6.17) and (6.18) imply

$$z^T z = (C_1 x + D_{12} u)^T (C_1 x + D_{12} u) = x^T C_1^T C_1 x + u^T D_{12}^T D_{12} u.$$

- If we want to find a solution such that

$$\|T_{zw}\|_\infty \leq \gamma$$

then, after some elementary computations, it is possible to show that the Riccati equation becomes

$$A^T X + X A + C_1^T C_1 + X \left(\frac{B_1 B_1^T}{\gamma^2} - B_2 B_2^T \right) X = 0 .$$

Note that when $\gamma \Rightarrow \infty$ the H^∞ Riccati equation becomes the H^2 (LQ) Riccati equation with R=I

$$A^T X + X A + C_1^T C_1 - X B_2 B_2^T X = 0.$$

For more details on the 1-Riccati equation solution see Petersen [32].

6.3 H$^{\infty}$ CONTROL WITH OUTPUT FEEDBACK: 2-RICCATI EQUATION SOLUTION

We outline next the 2-Riccati approach to H$^{\infty}$ control developed in Doyle, Glover, Khargonekar and Francis [20].

Definition 6.3 [20]

Consider the Riccati equation

$$A^T X + X A + X R X - Q = 0$$

where X, R, Q are real symmetric n x n matrices.

The associated Hamiltonian is

$$H = \begin{bmatrix} A & R \\ Q & -A^T \end{bmatrix} \tag{6.21}$$

If we now assume that H has no imaginary eigenvalues (which is referred to as the stability property) then it must have n eigenvalues in the LHP and n in the RHP.

Construct a matrix with columns composed of the eigenvectors corresponding to those eigenvalues in LHP and partition this matrix as

$$\begin{pmatrix} X_1 \\ X_2 \end{pmatrix}$$

where $X_1, X_2 \in \mathbb{R}^{n \times n}$.

If X_1 is a nonsingular matrix(which is referred to as the complementarity property) then

$$X = X_2 X_1^{-1} \tag{6.22}$$

is a solution to the Riccati equation.

In this case we denote this solution as

$$X = \text{Ric}(H) \tag{6.23}$$

and denote the domain of all the Hamiltonian matrices H satisfying the stability and the complementarity property as dom(Ric).

Consider now the control feedback configuration of Fig.6.1

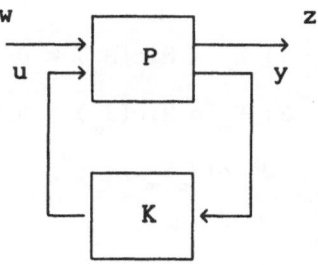

Fig. 6.1 Feedback system

where P is a linear system described by the state space equations

$$
\left\{
\begin{array}{l}
\dot{x} = A x + B_1 w + B_2 u \\
z = C_1 x + D_{12} u \\
y = C_2 x + D_{21} w
\end{array}
\right.
\tag{6.24}
$$

To simplify the formulation the following assumptions are made on P

$$
D_{12}^T \begin{bmatrix} C_1 & D_{12} \end{bmatrix} = \begin{bmatrix} 0 & I \end{bmatrix};
\tag{6.25}
$$

$$
\begin{bmatrix} B_1 \\ D_{21} \end{bmatrix} D_{21}^T = \begin{bmatrix} 0 \\ I \end{bmatrix};
\tag{6.26}
$$

A general solution without the special assumptions (6.25) and (6.26) can be found in [24].

Theorem 6.2 [20].

There exists a compensator K(s) such that

$$\|T_{zw}(s)\|_{\infty} < \gamma$$

if and only if

1) $X_{\infty} = Ric(H_{\infty}) \geq 0;$

2) $Y_{\infty} = Ric(J_{\infty}) \geq 0;$

3) $\rho(X_{\infty} Y_{\infty}) < \gamma^2;$

where

$$H_{\infty} = \left[\begin{array}{c|c} A & \dfrac{B_1 B_1^T}{\gamma^2} - B_2 B_2^T \\ \hline -C_1^T C_1 & -A^T \end{array} \right] ;$$

$$J_{\infty} = \left[\begin{array}{c|c} A^T & \dfrac{C_1^T C_1}{\gamma^2} - C_2^T C_2 \\ \hline -B_1 B_1^T & -A \end{array} \right] ;$$

and $\rho(\cdot)$ denotes the spectral radius of a matrix (the eigenvalue with maximum modulus).

When the conditions of the above theorem hold, a solution is

$$K(s) := \left[\begin{array}{c|c} \hat{A}_{\infty} & -Z_{\infty} L_{\infty} \\ \hline F_{\infty} & 0 \end{array} \right]$$

where

$$\hat{A}_{\infty} = A + \dfrac{B_1 B_1^T X_{\infty}}{\gamma^2} + B_2 F_{\infty} + Z_{\infty} L_{\infty} C_2 ;$$

$$L_{\infty} = -Y_{\infty} C_2^T; \qquad F_{\infty} = -B_2^T X_{\infty} ; \qquad Z_{\infty} = \left(I - \dfrac{Y_{\infty} X_{\infty}}{\gamma^2} \right)^{-1}.$$

All the other solutions can be parameterized by an H^∞ transfer matrix Q, with H^∞ norm bounded by γ,

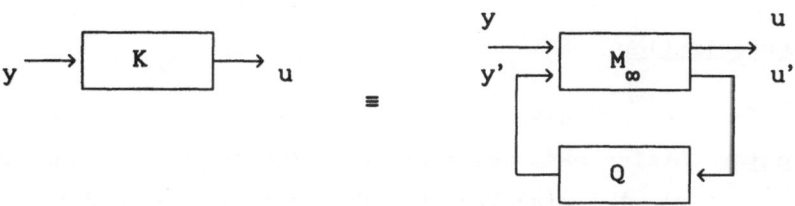

Fig 6.2 Parameterization of compensators.

where

$$Q \in M(H_\infty) ; \qquad \|Q(s)\|_\infty < \gamma ;$$

and

$$M_\infty(s) := \left(\begin{array}{c|cc} \hat{A}_\infty & -Z_\infty L_\infty & Z_\infty B_2 \\ \hline F_\infty & 0 & I \\ C_2 & I & 0 \end{array} \right) .$$

6.4 MATLAB™ : ROBUST CONTROL SOFTWARE

Many of the matrix computations required in chapter 5 and 6 can be done with the software package MATLAB™-Robust Control toolbox [9]. We describe here just a few of the relevant functions in MATLAB™.

(1) Q-parameterization

MATLAB™ function : **youla**

Input data: State-space realization (A, B, C, D) of the "augmented" plant ($p(s)$ transfer function in figure 6.1).

Output data: State-space realization of the transfer function between input w and output z (figure 6.1) when the compensator $K(s)$, is Q parameterized as in (5.2).

$$T_{zw}(s) = T_{11}(s) + T_{12}(s) \, Q(s) \, T_{21}(s)$$

$$T(s) = \begin{pmatrix} T_{11}(s) & T_{12}(s) \\ T_{21}(s) & 0 \end{pmatrix}.$$

Comment : This function allows for a direct computation of the transfer function in (5.1) without going through the separate computations of coprime matrix fractions or inner/outer factorizations.

(2) Balanced realization

MATLAB™ function : **obalreal**

Input data: A realization of the transfer function $G(s)$, i.e.

$$G(s) := \begin{bmatrix} A & B \\ C & D \end{bmatrix}.$$

Output data: A ordered balanced realization of $G(s)$, together with the vector g of diagonal elements of P and Q, i.e. $g = (\sigma_1, \sigma_2, \ldots, \sigma_n)$, and the similarity transformation t which convert the given realization to a ordered balanced realization.

Comment: This function can be used to compute the Hankel norm of G(s).

(3) Minimal Realization

MATLAB™ function : **minreal**
Input data: State-space realization, i.e. A,B,C,D matrices.
Output data: Minimal state space realization.
Comment: A number of results quoted in chapter 5 and 6 and functions in MATLAB™ require a minimal realization as input data.

(4) Lyapunov Equations

MATLAB™ function : **lyap**
Input data: Matrices A,B,C.
Output data: Solution X of the general Lyapunov equation, A X + X B = -C.
Comment: Actually **lyap** is a function in the **Control toolbox**, however the robust toolbox software requires the control toolbox software as a subset in order to run.

(5) Riccati equations

MATLAB™ function : **aresolv**
Input data: Matrices A, W, V.
Output data: Solution X to the algebraic Riccati equation,
$$A^{T}X + X A + W + X V X = 0$$
Comment: The function **aresolv** can be used to solve the H^{∞} state and output feedback control. The matrix V need not be negative definite as is normally required in optimal LQ problems.

(6) <u>Augmented plant</u>

MATLAB™ function : **augtf**

<u>Input data:</u> State-variable realization of the plant G(s), see figure 1.2, and transfer functions for "weighting" functions $W_1(s)$, $W_2(s)$ and $W_3(s)$ on the signal e(t), u(t) and y(t), respectively.

<u>Output data:</u> State-variable realization of the augmented plant P(s), see figure 6.1.

<u>Comment:</u> The output vector z in the augmented plant P(s) in figure 6.1 includes the effect of weighting functions.

(7) <u>Hankel Norm solution of the H^∞ control problem</u>

MATLAB™ function : **linf**

<u>Input data:</u> Minimal state-space realization of "augmented" plant P(s) (fig. 6.1).

<u>Output data:</u> State-space realization of the controller K(s), see fig. 6.1, which minimizes the H^∞ norm, $\|T_{zw}\|_\infty$.

<u>Comment:</u> The **linf** function may be used to solve sensitivity and disturbance rejection problems.

(8) <u>2-Riccati equation solution of the H^∞ control problem</u>

MATLAB™ function : **hinf**

<u>Input data:</u> State-space realization of "augmented" plant P(s) (fig. 6.1).

<u>Output data:</u> State-space realization of the "augmented" controller K(s), see fig. 6.2, which yields all compensators which guarantee the H^∞-norm bound $\|T_{zw}\|_\infty \le 1$.

<u>Comment:</u> The compensator produced by this function is parameterized by an arbitrary bounded-real matrix Q(s). See figure 6.2. Note that the notation in

function $M_\infty(s)$ in [20], shown in figure 6.2, and K(s) in [20] is F(s) in [9].

The following example illustrates a MATLAB™ computation.

Example 6.1

Given the augmented plant

$$P(s) := \left[\begin{array}{c|cc} A & B_1 & B_2 \\ \hline C_1 & D_{11} & D_{12} \\ C_2 & D_{21} & D_{22} \end{array}\right] = \left[\begin{array}{c|cc} 1 & 0 & -1 \\ \hline 0 & 0 & 0.2 \\ -1 & 1 & 0 \end{array}\right].$$

The MATLAB™ function **linf** generate the following compensator

$$F(s) := \left[\begin{array}{c|c} acp & bcp \\ \hline ccp & dcp \end{array}\right] := \left[\begin{array}{c|c} -3.381 & 1.543 \\ \hline -0.617 & 0 \end{array}\right];$$

or equivalently

$$F(s) = \frac{-0.9523}{s + 3.381} .$$

This compensator guarantees that

$$\|T_{zw}\|_\infty \le 1$$

where the signals z and w are defined in fig. 6.1 and equation (6.24).

Note: The augmented plant in example 6.1 correspond to a problem of additive robust stabilization for an open-loop plant with transfer function $G(s) = \frac{1}{1 - s}$ and an uncertainty bound function, r(s)=0.2. For this scalar problem, the interpolation approach in section 2.4 yields the following compensator, F(s)=-2.

CHAPTER 7
MULTIOBJECTIVE DESIGN

In this chapter we discuss some problems that involve more than just a single performance measure, i.e. robust stability, sensitivity minimization, etc, which is likely to be the situation in most practical problems. However we take the position that whatever additional performance measure is imposed, robust stability must be a minimal requirement for all systems. With this point of view we focus on the parameterization of all robustly stabilizing compensators. As has been shown in the previous chapters, for unstructured frequency domain perturbations, robustly stabilizing compensators can be parameterized in terms of arbitrary bounded real functions, and as arbitrary bounded real matrices in the multivariable case (as will be discussed in section 7.1). We call such bounded real functions, U parameters, in contrast to the Q parameter which is an arbitrary H^{∞} function, or matrix, which is used to guarantee only nominal stability. The literature is not consistent in the use of Q and U parameters, so that one must carefully note the properties of the particular function being cited. Since most multiobjective problems are multivariable, we present in section 7.1 some robust stability results for multivariable systems, which represent extensions of the single-input-single-output results presented in chapter 2.

7.1 MULTIVARIABLE ROBUST STABILIZATION

We paraphrase below some theorems on multivariable robust stabilization which appear in Vidyasagar [37], without proofs. Proofs may be found in [37].

Theorem 7.1 (Theorem 4 on page 273, [37])
Consider the perturbed plant given by (additive perturbation)

$$P(s) = P_0(s) + \delta P(s),$$

where

$$\|\delta P(j\omega)\| < |r_A(j\omega)|, \text{ all } \omega \tag{7.1}$$

where $r_A(s)$ is an outer function while $P(s)$ and $P_0(s)$ have the same number of RHP unstable poles; then a compensator $c(s)$ robustly stabilizes all permissible plants as defined above if and only if

$$\|c(s)(I + P_0(s)c(s))^{-1}r_A(s)\|_\infty \leq 1 \tag{7.2}$$

Note: The norms in (7.1) and (7.2) are defined as follows

$$\|A\| = \bar{\sigma}(A)$$

$$\|B(s)\|_\infty = \sup_\omega \bar{\sigma}(B(j\omega))$$

where $\bar{\sigma}(A)$ denotes the largest singular value of the matrix A.

Theorem 7.2 (Theorem 6 on page 273, [37])
Consider the perturbed plant given by (output multiplicative perturbation)

$$P(s) = (I + M(s))P_0(s);$$

where

$$\|M(j\omega)\| < |r_M(j\omega)|, \text{ all } \omega \tag{7.3}$$

where $r_M(s)$ is a function with all finite poles in Re $s<0$, then $c(s)$ robustly stabilizes all permissible plants as defined above if and only if

$$\left\| P_0(s)c(s)(I + P_0(s)c(s))^{-1}r_M(s) \right\|_\infty \leq 1.$$

We are using here the notation in [37]. The configuration then for the plant and compensator is the configuration in figure 2.1.

With the following definitions

$$R(s) = c(s)(I + P_0(s)c(s))^{-1}$$

$$S(s) = (I + P_0(s)c(s))^{-1} \text{ , sensitivity matrix}$$

$$T(s) = P_0(s)c(s)(I + P_0(s)c(s))^{-1}, \text{ complementary sensitivity matrix } (T=I-S)$$

we may define performance measures

$$\left\| W_1(s)S(s) \right\|_\infty \tag{7.4}$$

$$\left\| W_2(s)R(s) \right\|_\infty \tag{7.5}$$

$$\left\| W_3(s)T(s) \right\|_\infty \tag{7.6}$$

which can represent sensitivity, additive-perturbation stability, and multiplicative-perturbation stability design measures, where the weights W_1, W_2 and W_3 are selected appropriately for the problem at hand. For example for additive-perturbation stability $W_2(s)=r_A(s)$, and we then require $\left\| W_2(s)R(s) \right\|_\infty \leq 1$. It is clear from the above that the various performance measures can be viewed as H^∞ control problems. Finally from our previous discussions in chapters 2-6 it follows that all robustly stabilizing compensators can be parameterized in terms of a <u>bounded real matrix</u> (a matrix $U(s)$ which belongs to $M(H^\infty)$ with the additional property, $\left\| U(s) \right\|_\infty \leq 1$).

7.2 U-PARAMETER DESIGN

The basic idea behind U-parameter design [15] is to parameterize all robustly stabilizing compensator, either for additive or multiplicative perturbations, by a free bounded real matrix U(s), and then use this free parameter for further design. From MATLAB™ software one can compute the "augmented" compensator K(s), see figure 7.1,

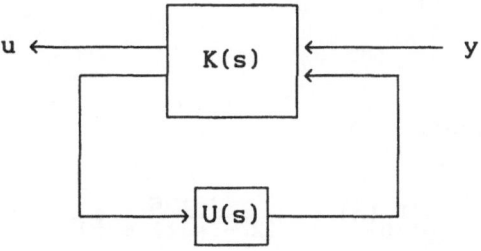

Fig. 7.1 Augmented compensator

which guarantees

$$\|r_{A}(s)\ R(s)\|_{\infty} \leq 1 \qquad (7.7)$$

for additive stability robustness or

$$\|r_{M}(s)\ T(s)\|_{\infty} \leq 1 \qquad (7.8)$$

for multiplicative stability robustness. Then, for example, if one wants to design a feedback system which is robustly stable respect to additive perturbations and have optimal disturbance rejection for the nominal system, one can do so by computing a K(s) which satisfies (7.7) and use the free bounded real matrix U(s) to minimize

$$\|W_{1}(s)\ S(s)\|_{\infty} \qquad (7.9)$$

where S(s) is given by $S(s)=(I+P_{0}(s)C(s))^{-1}$. Transfer functions such as $W_{1}(s)S(s)$ can be expressed in terms of U(s) as <u>linear</u>

fractional transformations, i.e.

$$W_1(s)S(s) = P_{11}(s) + P_{12}(s)U(s)\Big[I - P_{22}(s)U(s)\Big]^{-1} P_{21}(s). \qquad (7.10)$$

Unfortunately U(s) does not appear in an affine way in the transfer function (7.10), as does the Q-parameter discussed in chapter 5. However specific numerical problems can be solved by nonlinear programming techniques by further parameterization of U(s). We illustrate this U-parameter design approach by some simple scalar examples taken from [15].

Example 7.1
Consider a nominal plant

$$P_0(s) = \frac{1 - s}{(2 - s)(1 + s)}$$

with additive perturbation bounded by,

$$|\delta P(j\omega)| < \frac{1}{20}, \text{ i.e. } r_A(s) \equiv \frac{1}{20}.$$

The design objectives are to compute a compensator which guarantees robust stability for _perturbed plants_ and yields a minimal value of

$$\Phi = \|W(s)\ S(s)\|_2 = \left(\int_{-\infty}^{\infty} |W(j\omega)S(j\omega)|^2 d\omega \right)^{1/2}$$

for the _nominal plant_, for the weighting function $W(s) = \dfrac{1}{s + 1}$. Using the interpolation theory in section 3.3 all robustly stabilizing compensators may be parameterized as follows

$$C(s) = \frac{A_1(s)U(s) + A_2(s)}{A_3(s)U(s) + A_4(s)}$$

where

$$A_1(s) = 20\ s^2 - 20\ s - 40$$

$$A_2(s) = 12\ s^2 + 36\ s + 24$$

$$A_3(s) = 0.6\ s^2 - 18.2\ s + 21.2$$

$$A_4(s) = s^2 - 5\ s - 14$$

see exercise 3 in section 3.4.

For the simplest choice of U(s), i.e. U(s)=a, $|a| \leq 1$, the minimal value of "a" which minimizes Φ may be computed directly to be a=0.42, with a nominal value of Φ given by 8.1989. If Φ were minimized with no robust stability constraint then a minimal value of Φ=5.3147 can be achieved.   □

Example 7.2

Consider the same robust stabilization problem as in example 7.1, but with the additional requirement that the system have a zero steady-state response to a unit step and that the settling time for the nominal plant be minimized.

In order to meet the requirements in this case a more complex U(s) is required.

If we select

$$U(s) = \frac{a\ s + b}{s + 1}$$

then U(s) will be bounded real if $|a| \leq 1$ and $|b| \leq 1$. To meet the zero steady-state requirement we need b_1=0.66. The value of "a" is then used to minimize the settling time. By direct evaluation we obtain a minimal settling time (for 10 % of steady state) for a=0.75.

                                                      □

7.3 SURVEY OF OTHER APPROACHES TO MULTIOBJECTIVE DESIGN

In this section we review some other approaches to robust multiobjective design. A more complete collection of recent results in this area may be found in [17], section 2.2.

(1) <u>MATLAB^{TM} Approach</u>

A multiobjective design problem, such as for example, the problem of guaranteeing additive-perturbation robust stability while minimizing an H$^{\infty}$-norm on weighted sensitivity can be solved using MATLAB^{TM} software by defining the following performance measure

$$\left\| \begin{matrix} \gamma\, W_1(s)S(s) \\ W_2(s)R(s) \end{matrix} \right\|_{\infty} \le 1 \qquad (7.11)$$

and increasing the value of γ until the inequality (7.11) is just satisfied. This has the effect of satisfying the robust stability condition $\|W_2(s)\ R(s)\|_{\infty} \le 1$, when $W_2(s)=r_A(s)$, while making $\|W_1(s)\ S(s)\|_{\infty}$ as small as possible, since the satisfaction of (7.10) implies $\|W_2(s)\ R(s)\|_{\infty}$ and $\|W_1(s)\ S(s)\|_{\infty} \le 1/\gamma$.

The MATLAB^{TM} function **hinf** or **linf** may be used to find the compensator which satisfies (7.11). One disadvantage of this approach is the results are conservative since inequality (7.11) may be violated even though both terms in (7.11) may individually have norms less than 1. In contrast the U-parameter approach separately guarantees that, $\|W_2(s)\ R(s)\|_{\infty} \le 1$, while then minimizing $\|W_1(s)\ S(s)\|_{\infty}$. Another difficulty is that the order of the compensator is equal to the order of the plant plus the order of both weights $W_1(s)$ and $W_2(s)$. With U-parameter design one can start optimizing with compensators with order equal to the order of the plant plus only the order of $W_2(s)$.

(2) <u>Q-parameterization and convex programming approach [6]</u>

As noted in chapter 5, Q-parameterization may be used to generate transfer functions which are affine in Q, i.e. of the form

$$T_1 + T_2 Q T_3. \tag{7.12}$$

Since any norm on an affine in Q function is <u>convex</u> in Q, and inequality constraints of the form

$$\|T_A + T_B Q T_C\|_\infty \leq 1$$

define a convex constraint set for Q, problems involving both minimization of transfer functions and the satisfaction of H_∞-norm inequalities (robust stabilization conditions), can be reduced to <u>convex programming</u> problems. One important property of a convex nonlinear programming problem is that the solution is known to be <u>unique</u>.

The Q-parameter function Q(s) may be further parameterized in such a way to preserve convexity by choosing Q(s) of the form

$$Q(s) = \sum_{i=1}^{N} g_i Q_i(s) \tag{7.13}$$

where the $Q_i(s)$ are given H^∞ matrices and g_i are the new design parameters. The value of N is chosen to give a desired degree of accuracy for the minimal value.

The following example of the convex programming approach is taken from [30].

Example 7.3

Consider the problem of minimizing

$$\Phi = \left\| W_1(s) \, (I + P_0(s)C(s))^{-1} \right\|_\infty$$

subject to the multiplicative robust stability constraint

$$\Phi_3 = \left\| W_3(s)P_0(s)C(s) \, (I + P_0(s)C(s))^{-1} \right\|_\infty \leq 1$$

where

$$P_0(s) = \frac{s-2}{s-12} \; ; \; W_1(s) = \frac{1}{30}\left(\frac{s+6}{s+1}\right);$$

$$W_3(s) = \frac{1}{3}\left(\frac{s+1}{s+2}\right)\frac{(s+6)^2}{(s^2+2s+37)} \; .$$

The value of N=20 was selected with

$$Q_1(s) = \left(\frac{s-20}{s+20}\right)^1 .$$

The Q-parameterization for this problem is done with

$$P_0(s) = \frac{N(s)}{D(s)}$$

where

$$N(s) = \frac{s-2}{s+6} \quad \text{and} \quad D(s) = \frac{s-12}{s+6} \; .$$

Since Q(s) is of very higher order, since N=20, the resulting compensator is approximated by the following third order compensator

$$C(s) = \frac{-1.5345(s-0.084796)(s^2+1.5166s+35.658)}{(s+0.86685)(s^2+1.3714s+24.576)} \; .$$

With this third order compensator one obtains

$$\Phi_1 = 0.1995$$

and

$$\Phi_3 = 0.9929.$$

Using the U-parameter approach with U(s) = constant, one obtains

$$\Phi_1 = 0.2943$$

$$\Phi_3 = 0.9361$$

with a third-order compensator.

An important advantage of this approach, versus
U-parameterization, is that the minimization problem is convex.
One potential difficulty is that a high order compensator may be
required. However often, by simple model reduction methods,
reduced order compensator can be obtained, as in the above
example.

(3) <u>Coupled-Riccati-Equation approach [3]</u>

It is possible to show that the problem of optimal static output
feedback for the minimization of an H^2-norm on a given transfer
function while guaranteeing a fixed bound on an H^∞-norm of
another transfer function (which can represent a robust stability
constraint), can be reduced to the solution of two coupled
Riccati equations [3]. This result can obviously be used for
certain multiobjective design problems. At this point a major
practical obstacle to the application of this approach is the
lack of efficient numerical algorithms for the solution of the
coupled Riccati equations. One advantage, when solutions can be
obtained, is that the resulting compensator will be of low order
A major theoretical obstacle is that there are no conditions
known for the existence of solutions for fixed order
compensators.

7.4 GENERAL MULTIOBJECTIVE DESIGN

While a general theory does exists for the simultaneous "minimization" of an arbitrary set of scalar performance measures $f_i(K)$, $i=1,\ldots n$, where K is a design variable, see for example [34], analytical solutions are available in only limited cases. Multiobjective optimization of this type is sometimes referred to as "vector"optimization. The only meaningful "optimal" solutions of vector optimization problems of this type are the so called non-inferior (Pareto-optimal) solutions. A solution to vector optimization problem K^* is said to be non-inferior, or Pareto-optimal, if there exists no admissible K such that

$$f_i(K) \leq f_i(K^*), \text{ all } i$$

and

$$f_j(K) < f_j(K^*)$$

for some j.

Even when non-inferior solutions can be found there is always the issue of tradeoffs between such solutions. In any case for robust system design, robust stability conditions must always be given priority over all other performance measures.

We quote next a commonly used theorem in vector optimization [36].

<u>Theorem</u> : If the functions $f_i(K)$ are convex in K and K^* is Pareto-optimal, then there exists λ_i, where $\lambda_i \geq 0$ and

$$\sum_{i=1}^{n} \lambda_i = 1$$

such that K^* is a solution to the scalar optimization problem

$$\min_{K} \sum_{i=1}^{n} \lambda_i f_i(K) .$$

Conversely if given a set λ_i there is a unique solution to the above scalar optimization problem, K^*, then K^* is Pareto-optimal.

The above theorem is an example of <u>scalarization</u> results commonly used in vector optimization problems. Unfortunately the scalar optimization problem seldom has an analytic solution and some numerical or iterative approximation method is required to find Pareto optimal solutions. In [36] an iterative approximation method is used to solve a multiobjective problem with two objectives, one of which represents a robust stability constraint and the other a weighted nominal sensitivity design objective. Both objectives are taken to be H^∞ norms and a problem of this type is often referred to as the <u>two-disk</u> problem. The numerical example studied in [30] is the same as the one studied in [36]. In the following table we compare the results obtained by the two methods.

Method	$f_1(K)$	$f_2(K)$
Convex Programming [30]	0.1995	0.9929
Approximate Scalarization [36]	0.7998	0.9999

Multiobjective Design Example

In the above example the nominal plant is

$$P_0(s) = \frac{s - 2}{s - 12}$$

the plant perturbation is assumed to be multiplicative, i.e.

$$p = p_0(1 + e),$$

with

$$|e(j\omega)| < |W_2(j\omega)|$$

where

$$W_2(s) = \frac{1}{3} \left(\frac{s + 1}{s + 2} \right) \frac{(s + 6)^2}{s^2 + 2 s + 37} .$$

The design parameter is the compensator $K(s)$ and the design objective are to "minimize" the two functions

$f_1(K) = \|W_1(s) (1 + p_0(s) K(s))^{-1}\|_\infty$, weighted sensitivity, and

$f_2(K) = \|W_2(s) p_0(s) K(s) (1 + p_0(s) K(s))^{-1}\|_\infty$, robust stability,

where $W_1(s)$ is the sensitivity weight given in this example by

$$W_1(s) = \frac{1}{30} \left(\frac{s + 6}{s + 1} \right).$$

Note that for robust stability we require that $f_2(K) < 1$ in all cases.

For this example the convex programming solution produces a "superior" solution, since all its component objective function values are less than those of the approximate scalarization approach.

REFERENCES

[1] B.D.O. Anderson and J.B.Moore, "Optimal Control: linear quadratic methods", Prentice-Hall International, 1989.

[2] B. R. Barmish, "New tools for robustness analysis", *Proceedings of the 27th Conference on Decision and Control*, December 1988, pp. 1-6. Also in Dorato [17].

[3] D.S. Bernstein, W.M. Haddad and C.N. Nett, "Minimal complexity control law synthesis,Part 2 : Problem solution via H_2/H_∞ optimal static output feedback", *Proc. 1989 American Control Conf.*,pp. 2506-2511, 1989. Also in Dorato [17].

[4] H.S. Black, "Stabilized feedback amplifiers", U.S. Patent No.2,102,671, 1927.

[5] H.W. Bode, "Network Analysis and Feedback Amplifier Design". Princeton, NJ : Van Nostrand, 1945.

[6] S. Boyd, C. Barrat and S. Norman, "Linear controller design: limits of performance via convex optimization", *Proc. IEEE*, vol. 78, pp. 529-574, March 1990.

[7] B.C. Chang and J.B. Pearson Jr., "Optimal disturbance reduction in linear multivariable systems", *IEEE Trans. Automat. Contr.*, vol. AC-29, pp. 880-887, 1984. Also in Dorato [13].

[8] C.T. Chen, "Introduction to Linear System Theory", New York: Hott, Rinehart and Winston, 1970.

[9] R.Y.Chiang, M.G. Safonov, "MATLAB-Robust Control Toolbox User's Guide", The MathWorks Inc, South Natick, MA, 1988.

[10] S. Cusumano, K. Poolla and T. Ting, "On robust stabilization synthesis for plants with block structured modeling uncertainty", *Proc. 26 IEEE Conf. on Decision and Control*, Los Angeles, CA, pp. 423-428.

[11] P. Delsarte, Y. Genin, Y. Kamp, "On the role of the Nevanlinna-Pick problem in circuit and system theory", *International Journal on Circuit Theory and Applications*, vol. 9, pp. 177-187, 1981. Also in Dorato [13].

[12] C.A. Desoer,R.W. Liu, J. Murray and R. Saeks, " Feedback system design: The fractional representation approach to analysis and synthesis", *IEEE Trans. Automat. Contr.*, vol. AC-25, pp. 399-412, 1980. Also in Dorato [13].

[13] P.Dorato, "Robust Control", IEEE PRESS, New York, N.Y., 1987.

[14] P.Dorato, Y. Li, "A modification of the classical Nevanlinna-Pick interpolation algorithm with applications to robust stabilization", *IEEE Trans. Automat. Contr.*, vol. AC-31, pp. 645-648, 1986.

[15] P. Dorato and Yunzhi Li, "U-parameter design of robust single-input-single-output systems", *IEEE Trans on Automat. Contr.*, vol. 36, Sept., 1991.

[16] P. Dorato, H.B. Park, Y. Li, "An algorithm for interpolation with units in H^{∞}, with applications to feedback stabilization", *Automatica*, vol. 25, pp. 427-430, 1989.

[17] P.Dorato, R.K. Yedavalli, "Recent Advances in Robust Control", IEEE PRESS, New York, N.Y., 1990.

[18] J. Doyle, "Analysis of feedback systems with structured uncertainties", *IEE Proceedings*, pp. 242-250, 1982. Also in Dorato [13].

[19] J.C. Doyle, "Synthesis of robust controllers and filters", *Proc. 22nd IEEE Conf. on Decision and Control*, pp. 109-114, Dec. 1983. Also in Dorato [13].

[20] J.C. Doyle, K. Glover, P.P. Khargonekar, B.A. Francis "State-space solutions to standard H_2 and H_∞ control problems", *IEEE Trans. on Automat. Contr.*, vol. 34, pp. 831-847, 1989. Also in Dorato [17].

[21] J.C. Doyle and G. Stein, "Multivariable feedback design: Concepts for a classical/modern synthesis", *IEEE Trans. Automat Contr.*, vol. AC-26, pp. 4-16, 1981. Also in Dorato [13].

[22] B.A. Francis, "A Course in H_∞ Control Theory", Lecture Notes in Control and Information Sciences, Vol. 88 Springer-Verlag, Berlin 1987.

[23] K. Glover, "All optimal hankel-norm approximation of linear multivariable systems and their L^∞-error bounds", *International Journal on Control*, vol. 39, pp. 1115-1193, 1984. Also in Dorato [13].

[24] K. Glover, J.C. Doyle, "State-space formulae for all stabilizing controllers that satisfy an H_∞-norm bound and relations to risk sensitivity", *Systems & Control Letters*, pp. 167-172, 1988.

[25] T. Kailath, "Linear Systems", Englewood Cliffs, NJ: Prentice Hall, 1980.

[26] V.L. Kharitonov, "Asymptotic stability of an equilibrium position of linear differential equations", *Differential Equations*, vol. 13, pp. 1483-1485, 1979.

[27] H. Kimura, "Robust stabilizability for a class of transfer functions", *IEEE Trans. Automat. Contr.*, vol. AC-29, pp. 788-793, 1984. Also in Dorato [13].

[28] H. Kwakernaak and R. Sivan, "Linear Optimal Control Systems", New York, Wiley Interscience, 1972.

[29] C.N. Nett, C.A. Jacobson, M.J. Balas, "A connection between state-space and doubly coprime fractional representations", *IEEE Trans. on Automat. Contr.*, vol. AC-29, pp. 831-832, 1984. Also in Dorato [13].

[30] S.A. Norman and S.P. Boyd, "Numerical solution of a two-disk problem", *Proc. 1989 American Control Conf.*, pp. 1745-1747. Also in Dorato [17].

[31] H. Nyquist, "Regeneration theory", *Bell. Syst. Tech. J.*, vol. 11, pp. 126-147, 1932.

[32] I.R. Petersen, "Complete results for a class of state feedback disturbance attenuation problems", *IEEE Trans. on Automat. Contr.*, vol. 34, pp. 1196-1199, 1989. Also in Dorato [17].

[33] M.G. Safonov, M.S. Verma, "L^∞ optimization and Hankel approximation", *IEEE Trans. on Automat. Contr.*, vol. AC-30, pp. 279-280, 1985. Also in Dorato [13].

[34] M.E. Salukvadze, "Vector-valued optimization problems in control theory", New York/London: Academic Press, 1979.

[35] A. Tannenbaum, "Feedback stabilization of plants with uncertainty in the gain factor", *Int. J. of Control*, vol. 32, pp. 1-16, 1980.

[36] T. Ting and K. Poola, "Upper Bounds and approximate solutions for multidisk problems", *IEEE Trans on Automat. Contr.*, vol. AC-33, pp. 783-786, 1988.

[37] M. Vidyasagar, "Control System Synthesis: A Factorization Approach", MIT Press, Cambridge, MA, 1985.

[38] M. Vidyasagar and H. Kimura, "Robust controllers for uncertain linear multivariable systems", *Automatica*, vol. 22, pp. 85-94, 1986. Also in Dorato [13].

[39] M. Vidyasagar and N. Viswanadham, "Algebraic design techniques for reliable stabilization", *IEEE Trans. Automat. Contr.*, vol. AC-27, pp. 1085-1095, 1982. Also in Dorato [13].

[40] J.L. Walsh, "Interpolation and approximation by rational functions in the complex domain", American Math. Society, Colloquium Pubblications, vol. XX, Providence, RI, 1956.

[41] D.C. Youla, J.J. Bongiorno, C.N. Lu, "Single-loop feedback stabilization of linear multivariable dynamical plants", *Automatica*, vol. 10, pp. 159-173, 1974. Also in Dorato [13].

[42] D.C. Youla, H.A. Jabr, J.J. Bongiorno "Modern Wiener-Hopf design of optimal controllers-Part I: The single-input-output case", *IEEE Trans. Automat. Contr.*, vol. AC-21, pp. 3-13, 1976. Also in Dorato [13].

[43] D.C. Youla, H.A. Jabr, J.J. Bongiorno "Modern Wiener-Hopf design of optimal controllers-Part II: The multivariable case", *IEEE Trans. Automat. Contr.*, vol. AC-21, pp. 319-338, 1976. Also in Dorato [13].

[44] D.C. Youla and M. Saito, "Interpolation with positive real functions", *J. Franklin Inst.*, vol. 284, pp. 77-108, 1967.

[45] G. Zames, B.A. Francis, "Feedback, minimax sensitivity, and optimal robustness", *IEEE Trans. Automat. Contr.*, vol. AC-28, pp. 585-601, 1983. Also in Dorato [13].

SUBJECT INDEX

118

R

Riccati equation, 82, 89, 94
Right coprime factorization, 71
Robust control problem, 5

S

Scalarization, 108
Schur function, 7, 40
Sensitivity function, 20, 60, 71
Simultaneous stabilization, 27
Small gain theorem, 31
spectral radius, 91
Stability internal, 11, 20
Stability property, 89
State-feedback, 82
Strictly bounded real function, 7
Strictly positive real function, 9
Strong stabilization, 19, 22

T

Two-Disk problem, 108

U

U-parameterization, 100
Uncertainty frequency-domain, 3
Uncertainty time-domain, 2
Unit function, 8, 48

V

Vector optimization, 107

W

Weighting function 95, 99

Lecture Notes in Control and Information Sciences

Edited by M. Thoma and A. Wyner

Lecture Notes in Control and Information Sciences

Edited by M. Thoma and A. Wyner

Lecture Notes in Control and Information Sciences

Edited by M. Thoma and A. Wyner

Errata

Robust Control for Unstructured Perturbations-An Introduction

Page 6. $\|F(s)\|_\infty = \sup_\omega \|F(j\omega)\|$ should be $\|F(s)\|_\infty = \sup_\omega |F(j\omega)|$.

Page 36. $|M(j\omega)| < |r(j\omega)|$ should be $|M(j\omega)| \leq |r(j\omega)|$. Also right below

$$p_0(j\omega)c(j\omega) + 1 \neq 0 \;\; \forall \omega$$

add: and that $p_0(j\omega)c(j\omega)$ have the correct number of encirclements of the -1 point required by the Nyquist stability criterion.

Page 37. At the top of page: "(2.18)" should be "(2.25)".

Page 73. At the bottom of page: $I + cp$ and S should be $I + cp = (Qn_l + v_r)^{-1}d_r^{-1}$ and $S = d_r(Qn_l + v_r)$.

Page 74. Just below (5.3), "$T_1 = -d_r v_r$" should be "$T_1 = d_r v_r$".

Page 76. In equation (5.6), the H^∞ norm of a matrix has not been defined. For a stable matrix $M(s)$, it should be defined as follows:

$$\|M(s)\|_\infty = \sup_\omega \bar{\sigma}(M(j\omega))$$

where $\bar{\sigma}(M(j\omega))$ denotes the largest singular value of the matrix $M(j\omega)$
At the bottom of the page $\|T_{11}+\tilde{Q}\|_\infty$ should be defined as the L^∞ norm (See definition 5.4), since T_{11} is unstable.

Pages 77 and 78. All norms on these pages, for example the norm on the right side of (5.7), which are norms on unstable functions should be L^∞ norms.

Page 79. Bottom of page: Actually $T(s)$, given by $T = \theta_2^* T_1 \theta_3^*$ is not totally unstable (all poles in RHP) because of the stable term T_1. However by a partial fraction expansion one can define a modified T_1 which is totally unstable, as is done in [33].

Page 81. Just below

$$\bar{\sigma} = \sigma_1 \geq \sigma_2 \geq \ldots$$

add: With an ordered balanced realization the Hankel singular values are precisely the diagonal elements of P, or Q.

There could be some confusion here between Hankel singular values and regular singular values, since the same notation is used in both cases. Hopefully the context of the problem will help distinguish the two.

Page 83. Top of page: $I - G^T(j\omega)G(j\omega)$ should be $I - G^T(-j\omega)G(j\omega)$. Just below this equation: $\|G(j\omega)\|_\infty \leq 1$ should be simply $\|G(j\omega)\| \leq 1$.

Page 103. Middle of page: (7.10) implies $\|W_2(s)R(s)\|_\infty \leq 1$ and $\|W_1(s)S(s)\|_\infty \leq 1/\gamma$.

Page 104. Middle of page: "One important property of a convex nonlinear programming problem is that the solution is know to be unique", should more properly read "is that the minimal value is know to be unique".

Page 110. Reference [1]. The correct date of this reference is 1990.